MASTERING SPACE

For over two hundred years the domination of some countries by others has been intrinsic to international relations, with national economic and political strength viewed as essential to a nation's survival and global position.

Mastering Space identifies the essential features of this 'state-centredness' and suggests an optimistic alternative more in keeping with the contemporary post-Cold War climate. Drawing on recent geopolitical thinking, the authors claim that the dynamism of the international political economy has been obscured through excessive attention to the state as an unchanging actor. Dealing with such topical issues as Japan's rise to economic dominance and America's perceived decline, as well as the global impact of continued geographical change, the book discusses the role of geographical organization in the global political economy, and the impact of increasing economic globalization and political fragmentation in future international relations.

The authors identify the present time as crucial to the global political economy, and explore the possibilities of moving the world from mastering space to real reciprocity between peoples and places.

John Agnew is a Professor of Geography at the Maxwell School of Citizenship and Public Affairs, Syracuse University. **Stuart Corbridge** is a Lecturer in Geography at the University of Cambridge and a Fellow of Sidney Sussex College.

MASTERING SPACE

Hegemony, territory and international political economy

*John Agnew and
Stuart Corbridge*

London and New York

First published 1995
by Routledge
11 New Fetter Lane, London EC4P 4EE

Simultaneously published in the USA and Canada
by Routledge
29 West 35th Street, New York, NY 10001

Typeset in Garamond by
Ponting–Green Publishing Services, Chesham, Bucks
Printed in Great Britain by
TJ Press (Padstow) Ltd, Padstow, Cornwall

Printed on acid free paper

British Library Cataloguing in Publication Data
A catalogue record for this book is available from
the British Library

Library of Congress Cataloguing in Publication Data
Agnew, John A.
Mastering space: hegemony, territory and international
political economy/John Agnew and Stuart Corbridge.
p. cm.
Includes bibliographical references and index.
1. Geopolitics. 2. World politics. 3. International relations.
4. International economic relations.
I. Corbridge, Stuart. II. Title.
JCCC319.A45
320.1'2–dc20   94–22292

ISBN 0–415–09433–X (hbk)
ISBN 0–415–09434–8 (pbk)

CONTENTS

FIGURES

TABLES

PREFACE

In my opinion, the rights of man consist in the authorisation to take possession of all that is unoccupied and to defend all that has been so acquired

(Moser 1790, in Meinecke, Historism: The Rise of a New Historical Outlook 1972, 285)

the earth is in effect one world, in which empty, uninhabited spaces virtually do not exist. Just as none of us is outside or beyond geography, none of us is completely free from the struggle over geography. That struggle is complex and interesting because it is not only about soldiers and cannons but also about ideas, about forms, about images and imaginings

(Said, Culture and Imperialism 1993, 7)

Nately could scarcely believe his ears. He had never heard such shocking blasphemies before, and he wondered with instinctive logic why G-men did not appear to lock the traitorous old man up. 'America is not going to be destroyed!' he shouted passionately.
'Never?' prodded the old man softly.
'Well . . .' Nately faltered.
   The old man laughed indulgently, holding in check a deeper, more explosive delight. His goading remained gentle. 'Rome was destroyed, Greece was destroyed, Persia was destroyed, Spain was destroyed. All great countries are destroyed. Why not yours? How much longer do you think your own country will last? Forever? Keep in mind that the earth itself is destined to be destroyed by the sun in twenty-five million years or so.'
   Nately squirmed uncomfortably. 'Well, forever is a long time, I guess'

(Heller, Catch 22 1962 [1977 edn, 259])

News and history are written in the context of a dominant discourse – a narrative drama that sets the terms in which events are judged. Following the death of President Richard M. Nixon in April 1994, his foreign policy achievements were widely praised. Yet most eulogies missed one of the most enduring of Nixon's actions: the decision to abrogate the Bretton Woods monetary system in 1971. The reason lies in the fact that the Big Story in which Nixon's presidency was set was the Cold War; in this context what mattered most about Nixon's period in office was the American opening to China and US *détente* with the former Soviet Union. International economics was secondary. With the end of the Cold War the discourse emerging into dominance is one about a world that is becoming more and more economically interdependent. Whatever his other successes and failures, President Clinton's lobbying for NAFTA (the trade agreement with Mexico and Canada), and on behalf of global trade in the GATT (General Agreement on Tariffs and Trade), will loom large in any future account of his presidency.

This book is largely about how the conduct of international affairs can only be understood in the context of the Big Stories of particular historical periods. But, more especially, it is about how the *geographical* character of international affairs in these different periods offers a key to understanding their conduct. 'Terrestrial space' has become an important point of focus for scholars seeking to understand such contemporary changes as the restructuring of the modern world economy and the causes of the end of the Cold War.

Among the attempts at explicit incorporation of 'space' or 'geography' into social science in general and the study of international relations in particular could be included the following: the revival of fixed-form geopolitics (the direct impact of geographical location on state behaviour) in the writings of Gray (1988) and Collins (1986); the use of concepts of geographical core and periphery in the writings of world-system and dependency theorists; Harvey's (1990) argument for the spatial shifting of investment by business in response to declining rates of profit; the incorporation of geographical distance into quantitative models of interstate conflict (e.g. O'Loughlin 1986; Ward 1990); and the interpretation of claims about geostrategy and the spatial ordering of foreign policy decisions by political élites often called 'critical geopolitics' (e.g. Dalby 1990; Luke 1991).

This recent interest in space is surprising in that one of the main features of the present age is the speed or pace of change. Surely, space, or the presumed effect of geographical location and spatial setting on economic and political life, is fixed and, hence, of little use as a focus for understanding change? In fact, the production of space and how it is conceived can be used to convey the sense of how change is occurring. But this is so only if space is historicized; put in a historical context rather than seen as a permanent set of influences or fixed backdrop upon which history is inscribed.

This book is about using the angle of vision provided by a focus on the changing spatial organization of political–economic activity to offer an alternative approach to the field of international political economy. This field seeks to explain how political power affects economic outcomes and how economic forces limit political action within the world economy (see Crane and Amawi 1991; Biersteker 1993). While it incorporates many of the interests and perspectives of the longer established field of international relations, it rejects the often singular concentration on political and military power that has characterized that enterprise. It also ranges well beyond an international level of analysis to global and local scales. It is in this respect that a geographical perspective, sensitive to questions of geographical scale and spatial representation, has particular merit.

This is conceptually challenging, however, because established theoretical perspectives, while making key assumptions about the nature of terrestrial space such as its 'natural' division into blocks of state-territorial space, have been largely unconscious of their spatial character or of the limited relevance of their spatial assumptions in different historical epochs. Typically, space is regarded as an unchanging 'essence' that constantly produces the same effects.

The idea for this book came from a paper we co-authored several years ago (Agnew and Corbridge 1989). This attempted to move discussion of 'geopolitics' away from the fixed effects of a limiting or determining global physical geography to an understanding of geographies as socially constructed in different historical epochs. This perspective was inspired by the difficulty of understanding in conventional terms the changes in the international political economy underway in the 1980s. In the modern world, a world created through the conquests and 'struggle over geography' of the past several hundred years, geopolitics is 'an active process of constituting the world order rather than an accounting of permanent geographical constraints. The 1980s [were] a time of crisis in geopolitics precisely because as an old order [was] dying a new one [had] not yet been born' (Agnew and Corbridge 1989, 267).

Mastering Space is a metaphor for the character of the international political economy of the past two hundred years. It refers to the intrinsically geographical processes of disciplining, subjugating, exploiting and developing places that have gone on in different ways during this period. The first part of the book provides a framework for understanding the mastering of space, an appropriately masculinist metaphor for the process it represents. But it also has a second meaning, that of *understanding* the role of space in international political economy, which alternative locutions such as 'dominating space' would miss. The second part considers the 'present crisis' in the global political economy and relates it to various symptoms of a growing deterritorialization of economic and political powers. Finally, a third section offers some thoughts on how the world can move beyond mastery to real reciprocity.

Over the past several years the work of the following people has had a tremendous influence on the thinking that went into this book: Hayward Alker, Mark Bassin, Tom Biersteker, Christopher Clapham, Robert Cox, Simon Dalby, Daniel Deudney, Bud Duval, Anthony Giddens, Stephen Gill,

Robert Gilpin, David Harvey, Geoffrey Hawthorn, Sudipta Kaviraj, Gerry Kearns, Paul Krugman, Charles Lipson, Timothy Luke, Doreen Massey, Joseph Nye, John O'Loughlin, Gearoid O'Tuathail, Edward Said, Duncan Snidal, Susan Strange, Peter Taylor, R. B. J. Walker, Alex Wendt and Eric Wolf. Our colleagues and friends Chris Bramall, Carlo Brusa, James Duncan, Alan Hudson, Naeem Inayatullah, Sarah Jewitt, Lou Kriesberg, Linda McDowell, Ron Martin, Donald Meinig, John Nagle, Gearoid O'Tuathail, Mauro Palumbo, Stephen Rosow, Mark Rupert, Joanne Sharp, Graham Smith and Jock Wills have been consistently supportive and helpful. Of course none of those named bears responsibility for what we have written. A version of Chapter 4 previously appeared in the *Review of International Political Economy* 1:1 (1994): 53–80.

We dedicate this book to our daughters (Katie, Christine and Joanne) in the hope of a world in which real reciprocity between people and places, and not mastery of some by others, will increasingly characterize global space.

John Agnew   Stuart Corbridge
Syracuse NY Cambridge, England
1 May 1994

ACKNOWLEDGEMENTS

The authors gratefully acknowledge permission to reproduce copyright material from the following: Figure 2.1 is reproduced with permission of St Martin's Press from data supplied by the Conflict and Peace Data Bank, University of Pittsburgh; Figure 2.2 was first published in J. Nijman (1992) 'The limits of superpower: the United States and the Soviet Union since World War II', *Annals of the Association of American Geographers* 82: 681–95; Figure 3.2 is reproduced from J. O'Loughlin and H. Van der Wusten (1990) 'Political geography of panregions', *Geographical Review* 80: 3, from an original drawing in *Facts in Review* (German Embassy, Washington DC, 10 April 1941); Figure 3.5 is reproduced from C. E. Pletsch (1981) 'The Three Worlds, or the division of social scientific labor, circa 1950–75', *Comparative Studies in Society and History* 23: 565–90, by permission of Cambridge University Press; Figure 7.2 is reproduced from P. Dicken (1992) *Global Shift: The Internationalization of Economic Activity* (2nd edn) by permission of Paul Chapman Publishing Ltd, London.

Every effort has been made to obtain permission to reproduce copyright material. If any proper acknowledgement has not been made, we would invite copyright holders to inform us of the oversight.

1

INTRODUCTION

The term 'geopolitics' has been popular at times of dramatic global political change and then has tended to recede from use. The term was first used by the Swede Rudolf Kjellen in 1899. It became associated with the formal model of geographical influences on global conflict proposed by the British geographer Halford Mackinder in the early twentieth century in his efforts to promote the field of geography as an aid to the practice of British statecraft. During the 1920s and 1930s Mackinder's model of a Eurasian 'heartland' rising to global dominance if not held in check by cohesive reaction from the (British-dominated) 'outer or insular crescent', was adopted by certain German geographers who used it to justify Nazi expansionist designs on Eastern Europe. Not surprisingly, the term suffered from guilt by association. After the Second World War the term 'geopolitics' fell into disuse because of its Nazi connotations and its reliance on ideas of environmental determinism from which professional geographers were in retreat.

In recent years geopolitics has undergone another revival, but this time with little agreement as to its precise meaning and influence. Contemporary usage ranges from classical concepts of seapower versus landpower in the distribution of power among states, to the ways in which political leaders name places as more or less important strategically, organize foreign policy accordingly, and operate militarily. A more anodyne formulation refers to geopolitics as the equivalent of political geography, in the sense of a real variation in political phenomena at all scales, including the global. The range of usage illustrates an important feature of the contemporary world situation: the collapse of agreement about terms that previously had meanings that (seemingly) were self-evident. It is in this context that we are proposing a 'new geopolitics'. The world has changed significantly over the past twenty years. We believe that this change has been so profound that it is necessary to change the way in which we think about the geography of international relations.

We believe that in redirecting the orientation of geopolitics we can also offer an important new perspective on the field of international relations in general and international political economy in particular. One of the boasts often made by the mainstream social science that was institutionalized in

Europe and North America during the course of this century is that its claims about 'order' in human society held for all times and in all places. In no field has this point of view been more popular than 'international relations'. One version of the conventional wisdom is put forward with his usual directness by Samuel Huntington. Referring to the need for the United States to remain the most powerful state in the world lest other less benign states assume the mantle of leadership, he claims, matter-of-factly, that 'No reason exists to assume that what has been true for millennia will cease to be true in the next hundred years' (Huntington 1993a, 70). In direct counterpoint to this claim, this book is designed to provide an historical–geographical introduction to international political economy.

INTERNATIONAL RELATIONS
AND INTERNATIONAL POLITICAL ECONOMY

During the 1950s and 1960s international relations as a field of study was concerned primarily with security, or the military threat posed by states to one another's territories. The Cold War between the United States and the Soviet Union dominated the writings of intellectuals and the agendas of political leaders, particularly in the United States where international relations was most strongly entrenched as a field of academic study. International economic issues attracted little sustained attention, except insofar as they engaged with questions of military policy or alliance politics.

Since the 1970s this has changed. Many new developments, such as the growth of the Organization of Petroleum Exporting Countries (OPEC), the demands of poor countries for a New International Economic Order, the abrogation by the US government of the post-war international monetary system based on the 1944 Bretton Woods Agreement, and the international debt crisis, have stimulated a sense that the study of international relations was sorely lacking in its ability to address fundamental 'global' issues. The field remained locked into perspectives on strategic security and international organization that were incapable of addressing the 'new' questions of the day. For its part, mainstream economics provided models of trade and money that had little relevance for understanding the political determinants of international economic relations. If the world of international relations was peopled mainly by hard-boiled, male Machiavellians, the dismal world of economics was increasingly populated by the 'social morons' so devastatingly lampooned by Amartya Sen (Sen 1978).

The end of the Cold War has reinforced the sense of a need for new perspectives on international relations. Not only have old enemies suddenly been transformed into supplicants, but a new world economy is now in the making. A market-based world economy that in the mid-1970s involved only two-thirds of the total world population at best – including some African countries on the margins of involvement – has since gained more than two billion new members with the opening up of China, the former Soviet Union and the countries of Eastern Europe, and the decline of protectionism in India and Latin America. If the 1890s was the decade when almost the whole world was finally divided into states and colonial territories, the 1990s promises to be the decade when almost the whole world is finally incorporated into the modern world economy. In this context it is not surprising that no consensus has yet emerged about how the workings of the international political economy should be explained.

SPACE AND GEOPOLITICS

There is now something of an impasse in international political economy. The main perspectives are imprisoned, in our view, by a conception of space and 'geography' that emphasizes fixity over fluidity, stasis over change. In this book we want to outline an alternative that breaks with all of them in identifying the centrality of a dynamic geopolitics to the working of the international political economy.

In its most common usage geopolitics refers to a fixed and objective geography constraining and directing the activities of states. For example, fixed geographical features of the world, such as the disposition of states in relation to the distribution of the continents and oceans, or fixed processes of territorial–economic expansion relative to military strength, are seen as determining or strongly conditioning the strategic possibilities and limits of particular states. This is the geopolitics of Kjellan and Mackinder – for whom 'man and not nature initiates [here is the domain of statecraft]; but nature in large measure controls' (Mackinder 1904) – and of Spykman, who declared that 'Geography is the most fundamental factor in the foreign policy of states because it is the most permanent. Ministers come and go, even dictators die, but mountain ranges stand unperturbed' (Spykman 1942, 41). Such fixed-form views of geopolitics have been joined in popularity by more recent attempts to define geopolitics in terms of fixed processes of one kind or another. Adding geographical distance to models of international conflict has been one such strategy and its logic is clearly displayed in various containment theories and models of imperial overstretch. Another strategy makes reference to the workings of some constant underlying logic to the 'long-waves' (or cycles) in the modern world system.

The two assumptions common to most of these perspectives are that power is (a) an unvarying ability to make others do your will resulting from advantages of geographical location, population and natural resources; and (b) an attribute of territorial states that attempt to monopolize it in competition with other states. Each of these assumptions is problematic.

Today, as at certain times in the past (for example, the period between 1500 and 1700 in Western Europe), relative economic power has begun to displace military force and conquest as an important feature of international relations.

It is no longer just a means to the end of military power but an end in itself. Technology, education and economic growth have become more important than conventional geopolitical attributes of power in determining relative success in the international system. The links between economic expansion, military power and political empire upon which the classic accounts of inter-state conflict rely have largely been cut. Possession of military superiority no longer translates automatically into enhanced political status. Contemporary Russia and Ukraine provide ample evidence of this.

In addition, more intangible aspects of power have become more obviously important. Relative economic vulnerability affects the conduct and outcome of bargaining between states independently of the absolute resources they have. Setting the agenda of international relations by imposing your ideas or shaping others' preferences – what Nye (1990b, 181) calls 'co-optive power' – can be done even if a country's military or 'command' power does not outweigh ·that of other states. Much of Britain's past influence and contemporary American power has rested upon the ability to persuade other states to go along with their agendas for organizing and policing the international political economy. As Nye reminds us, power is not a singular entity; it comes in several forms (including co-optive power and command power) and it has been associated with different leading states at different times in different combinations (see Table 1.1).

The work of Nye and others has helped to shape our view of geopolitics, which is quite different to the fixed-form and fixed-process views we have outlined. By geopolitics we understand the division of global space by institutions (states, firms, social movements, international organizations, armed forces, terrorist groups, etc.) into discrete territories and spheres of

Table 1.1 Leading states and types of power resources, fifteenth to twentieth centuries

Period	Leading state	Major resources
Sixteenth century	Spain	Gold bullion, colonial trade, mercenary armies, dynastic ties
Seventeenth century	Netherlands	Trade, capital markets, navy
Eighteenth century	France	Population, rural industry, public administration, army
Nineteenth century	Britain	Industry, political cohesion, finance and credit, navy, liberal norms, island location (easy to defend)
Twentieth century	United States	Economic scale, scientific and technical leadership, universalistic culture, military forces and alliances, liberal international regimes, hub of transnational communication

Source: Nye 1990b, 180

political–economic influence through which the international political economy is regulated materially and represented intellectually as a natural order
of 'developed' and 'underdeveloped', 'friendly' and 'threatening' areas. It is
that set of socially constructed, rather than naturally given, practices and ideas
through which the international political economy is realized geographically.
In this perspective even the 'anarchic' world of competitive territorial states
pictured in many varieties of international relations theory is the historical
outcome of the mutual recognition of sovereignty among states. It is the
dominance of this process of territorial identity and interest formation that
in certain historical circumstances has 'disempowered' non-state actors and
privileged interaction between states. There is nothing natural about a world
simply divided up into territorial states and their interactions with one
another. We share with an emerging school of 'critical geopolitics' the view
that geopolitics is implicit in both the practice of and writing about all
types of international relations; it must not be confined to a reading of a
world ordered geographically into a more or less fixed hierarchy of states,
cores and peripheries, spheres of influence, flashpoints, buffer zones and
strategic relations.

We will come back to critical geopolitics shortly. For the moment let us be
clear that four main theoretical presuppositions inform our understanding of
this 'new' geopolitics, and distance it in various ways from most other
theoretical positions in international political economy.

First, **the primacy of 'the' territorial state is not a trans-historical given,
but is specific to different historical epochs and different world regions**.
States differ historically and geographically in their external powers and in
their ability to regulate their own territories. More generally, there have been
historical periods, and there are world regions, in which the 'sovereignty' of
any one state in relation to other states and non-state organizational actors is
far from complete. This is apparent throughout the modern world economy
with regard to the determination of exchange and interest rates; many 'state'
powers have been ceded to the markets. It is also apparent in large parts of
the ex-colonial world, where the imposition of World Bank and IMF-inspired
structural adjustment programmes illustrates the limits of sovereignty in a so-
called world of 'nation-states'. The contemporary world economy is shaped
by pension funds and transnational companies, as well as by international
organizations and (often rather fragile) 'nation-states' (consider the fate of
Yugoslavia and the status of various 'quasi-states' in Africa: Jackson 1990). It
is a world economy marked above all by a globalization of production,
exchange and information flows which has brought with it not so much
spatial homogenization as a new round of geographical differentiation and
uneven development at all spatial scales. Against this background the territorial state is losing its geographical primacy.

Second, the forces driving the international political economy are by no
means static and unchanging. Nor do they derive clearly from some hidden

or essential inner dynamic. Historic shifts back and forth from territorial (autarkic) to interactional (open) strategies of economic development have long been apparent and deserve our closest attention. **They demand an approach to geopolitics – or geopolitical economy – that is diachronic (or historical in its conceptions of causation in global affairs) rather than synchronic (cross-sectional or timeless).** Nowhere is this clearer than in the case of war, the favourite topic of a more traditional international relations community. The classical theorists invoked to legitimize the long-standing nature of territorial-state centrism – Thucydides, Machiavelli, Hobbes and Clausewitz – can be given either 'alternative readings' or dismissed in the light of present trends. Clausewitz, for example, is famous for his statement that war is always subordinate to political will. But much recent warfare around the world has no connection to politics or the state in Clausewitz's sense, or to disciplined armies and high-tech weapons. Bandits in Somalia, genocidal murderers in Cambodia and Bosnia, and murderous children in Liberia, represent the movement of war beyond the mobilizing spaces of the territorial state into a do-it-yourself mode whose only requirements are a rifle, a death-wish and a hatred of some group that is readily targeted. Even the new high-tech warfare exemplified by the Gulf War in 1990–1, though ostensibly about the violation of the state boundaries of Kuwait, appeared to many spectators as a contest in video-space on behalf of a state whose assets were largely held outside the country, that was financed by an international consortium, and that was toiled for by hundreds of thousands of labourers imported from South Asia and elsewhere. The warfare itself excited more interest (in some places) than the putative causes advanced for why the war was necessary. In the USA the war almost became one of the sports metaphors used by military experts to describe it. The venerable distinctions between politics and war, and civil (or internal) and international war lose much of their intellectual meaning in a new context of DIY warfare and New World Order sport-war. As the world map recasts itself at some remove from the stable spaces of nation-state identities we can either pretend that nothing really important has happened, or we can begin to think diachronically about the geopolitical workings of a dynamic international political economy.

Third, **the relative success or failure of different localities and regions in the international political economy at any particular time is due to their historical accumulation of assets and liabilities and their ability to adapt to changing circumstances, and not the result of 'natural' resource endowments.** The recent rise of a resource-poor Japan is testimony to this view. It also confirms that such geographical 'path-dependence' as there is is subject to reversal – as witness the relative economic rise of Germany and Japan post-1945 when compared to the relative decline of a 'victorious' United Kingdom. At the level of the international system, too, there have been various geopolitical orders, including the Cold War Geopolitical Order that lasted from about 1945 to 1990, that have been constructed by thinking political actors and not just called into being by 'natural' or evolutionary factors.

Fourth, and finally, along with the changing ways in which the international political economy operates (new patterns of flows, transfers and interactions) come new representations of the division and patterning of global space. Lefebvre (1991) makes an analytical distinction between 'spatial practices', 'representations of space' and 'representational spaces' that is helpful in understanding the nature of the connections between geopolitical orders and the discursive representations of space implicit in the practices of foreign policy. Spatial practices refer to the material and physical flows, interactions, and movements that occur in and across space as fundamental features of economic production and social reproduction. Representations of space involve all of the concepts, naming practices, and geographical codes used to talk about and understand spatial practices. Representational spaces are the scenarios for future spatial practices or 'imagined geographies' that inspire changes in the representation of space with an eye to the transformation of spatial practices. We see these three sets of ideas and practices as dialectically related. Each is implicated in the nature of the other, so that no one concept can be given causal primacy over the others. For example, after the Second World War the US government could have continued to regard the Soviet Union as the ally it had been during the war (representation of space). But a decision was made to represent the erstwhile ally as a threat to the spatial practices of the 'free' world economy (representational space). Now the relatively stable representations of the Cold War are up for grabs again, and with them comes the possibility of new spatial practices that will have to have regard for a new era of globalization and deterritorialization. Our point, of course, is that as the 'production of international political–economic space' changes, so do the dominant models of how it works and, also, visions of alternatives to it. **A critical geopolitics refers, then, not only to the material spatial practices through which the international political economy is constituted, but also to the ways in which it is represented and contested.** Successive discourses of geopolitics take shape – sometimes uneasily and always unevenly – against the backdrop of an international political economy experiencing periodic crisis and restructuring.

THE ORGANIZATION OF THE BOOK

The rest of this book is written with every regard for these four theoretical presuppositions, but it is not organized in such a way that successive chapters are given over to each assumption in turn; the presuppositions confront the text at all points along the way from here to the Conclusion, and in each of the three Parts that follow.

In Part I we reflect further on some of the organizing concepts that structure both our view of geopolitics and the traditions of geopolitics we are concerned to critique. Chapter 2 is focused on questions relating to the historical geography of geopolitical orders. In the first part of this chapter we examine various spatial ontologies that are apparent in existing conceptions of order, dominance and hegemony in the field of international relations. We also outline and defend our own, more Gramscian, ontology of spatial order. In the second part of the chapter we present a narrative account of the three geopolitical orders that we believe best correspond to our spatial ontologies of order and hegemony. These are (1) The Concert of Europe – British Geopolitical Order, c.1815–75; (2) The Geopolitical Order of Inter-Imperial Rivalry, c.1875–1945; and (3) The Cold War Geopolitical Order, c.1945–90. We hope this narrative account is illuminated by the theoretical exegesis in the first part of the chapter. *Mastering Space* is not intended to serve as a comprehensive guide to the political geography of the modern world system (see Taylor 1993b), but we do recognize that our theoretical statements about international political economy (and geopolitics) must have recourse to a competent historical geography of decisive events and issues in selected regions of the world.

Chapter 3 is concerned with geopolitical discourse and takes up several of the themes that are central to an emerging 'critical geopolitics'. We aim to show how each of the geopolitical orders identified in Chapter 2 has been associated with a set of rhetorical understandings and dominant meanings through which a given geopolitical order has been realized in strategic and economic policies. We also show how an apparently diverse set of geo-political discourses has for many centuries been underpinned by a set of assumptions which in some cases reach back as far as Renaissance Europe. What might be called the vocabulary of the Great Powers has been shaped (successively) by a belief in the civilizing mission of the great and the good, by an appeal to a naturalizing biology of power (the fit and the unfit), and with reference to the modern ideologies of market capitalism and state socialism.

Chapter 4 builds on Chapters 2 and 3 to mount a sustained attack upon the privileging of the territorial state in much of the literature on international political economy. We argue that a changing global economic geography is exploding the fixity of the territorial state and is thereby creating a trap for those who want to build timeless models upon rapidly shifting foundations. (The fact that territorial states might yet reclaim some of their powers is not at odds with the broader argument we are making.)

Part II of the book has a more sustained emphasis. It examines the implications for the international political economy of deterritorialization in the contemporary period – a period identified as a time of 'geopolitical disorder'. Chapter 5 reaches back to our accounts of order and hegemony in Chapter 2 to examine whether the USA is in 'relative decline', as many would argue, or is still 'bound to lead'. If our conclusions are ambivalent it is because we challenge the terms of reference of this debate. The territorial US economy has done rather badly in recent years; in contrast, US-based companies have continued to do rather well. The concept of 'hegemony' is not a straight-forward one. Chapter 6 approaches the same issue from another angle, examining the rise to 'power' of such countries and regions as Germany, Japan, China and the European Community (European Union). Once again, our intention is not only to provide an empirical account of events and trends, but also to offer a theoretically informed investigation of the significance of these developments. We also show why Japan, in particular, should not neatly be slotted into a model of the rise and fall of Great Powers that has much in common with the model of apostolic succession. Such models misconceive the nature of dominance and hegemony in the contemporary geopolitical economy. This theme is elaborated on at length in Chapter 7, where we show that economic power can no longer be seen only as an attribute of states that have more or less of it. This is to fall into the territorial trap. The modern world system is characterized rather by a global political economy in which no single national economy has a dominating role and in which many actors, including multinational corporations and transnational banks, are global in outlook and organization. States must now operate in a global context in which they are effectively, but unevenly, internationalized. In this context, regions and localities within states have become increasingly vulnerable to pressures emanating from outside the 'national space'. Some regions and localities can also take advantage of these new 'spaces of flows' to join a global economic system more and more akin to a gigantic hierarchy of cities and city-states, and less and less akin to the hierarchy of 'nation-states' still described in the international relations literature, and still mapped out in the annual reports of the World Bank, the IMF and the United Nations. We examine the implications of this new conception of 'hierarchy' for existing accounts of 'development' in the so-called Third World.

Part III of the book comprises a concluding chapter (Chapter 8), that looks at the possibilities for 'order' and 'real reciprocity' in a world of porous (and sometimes fragmenting) territorial states. Although the chapter begins with some dark thoughts on the 'new world disorder' which seems to be taking shape today, it closes with a more optimistic reading of the possibilities for empowerment that are latent in a deterritorializing world system. This takes us back to the article we published in 1989. That article concluded by suggesting that 'The best we can hope for is the construction of *a geopolitical and economic order* which recognizes a dispersed-power principle appropri-ate to an age of complex interdependence, and which promotes a decen-tralized system of governance' (Agnew and Corbridge 1989, 284; emphasis in the original). In a still broader sense we would say that this book attempts to define a new meta-discourse about geopolitics that establishes the pos-sibility of a world order without recourse to the fixed determinations that have characterized most geopolitical discourses in the past. The great Arab philosopher, Ibn Khaldun (1967 edition), grasped the essential point about the historical constitution of spatial organization, and why it is so difficult to understand, when he wrote that 'When there is a general change of conditions, it is as if the entire creation had changed and the whole world had been altered, as if it were a new and repeated creation, a world brought into existence anew' (Ibn Khaldun 1967, 119).

Part I

Mastering Space

2

GEOPOLITICAL ORDER

In Chapter 1 we argued that the international political economy is best viewed through a historical–geographical lens. The purpose of this chapter is to engage in an analysis of the period 1815–1990 through identifying the essential geographical and other features of each of three global geopolitical orders which prevailed during different parts of this period. Most of the chapter is taken up with a narrative account of these geopolitical orders. However, before embarking on this task, some preliminary discussion of the spatial ontology and view of 'order' implicit in the argument of the chapter is provided. Ontology refers to the entities and processes invoked in an explanation. Typically, the spatial content of their ontologies has been largely neglected by the conventional theories of international political economy. The principal purpose of this chapter is to centre discussion of the evolution of the international political economy since 1815 around the topic of spatial ontology.

SPATIAL ONTOLOGY

Spatial terms such as core and periphery, territory and location have become increasingly popular as locutions in the social sciences and humanities. By and large, however, the use of geographical terminology has been metaphorical and descriptive rather than ontological. 'Regional' and 'local' become synonyms for the 'particular' while 'global' signifies the 'general' or 'universal'. 'Territory' and 'location' evoke abstract images of power and wealth differentials without reference to any geographical basis to these differences themselves. Places just *happen* to differ; difference itself is constituted by abstract forces of class, ethnicity or gender. Cultural or economic differences are still thought of as produced extra-geographically by abstract 'forces' of class, ethnicity, or discursive hegemony, even as they are captured in geographical metaphors or acknowledged as manifesting themselves geographically. Geography may 'matter', but only as the *moment* in which abstract universal social processes, such as social stratification, state-building, and ideological hegemony, are revealed in space. Even for some who claim that 'space matters', therefore, it matters only contingently rather than necessarily (Malmberg 1992).

This implies that it is possible to formulate abstract theories which identify the necessary relations between logical objects (states, classes, groups, etc.). Only in concrete empirical research is there a need to account for the contingencies of spatial arrangements. The theories remain unaffected. If the human world was closed and unchanging this would be unproblematic. Spatial effects would be predictable. The problem is that the human world is open and changing and the precise relationship between objects and the spatial fields that contain them is dynamic. To understand the world, therefore, requires that we understand its changing geography (Agnew 1989).

Typically, however, the abstract concepts that are seen as producing political–economic difference have embedded in them implicit assumptions about how space mediates social processes (Agnew 1993). The division of the world into territorial entities we call 'states' produces actors that operate on a territorial definition of space i.e. a world divided into discrete and mutually exclusive blocks of space. Rather than a natural and universal process, however, this type of geographical division has a clear, if largely unexamined, historicity. It originated in seventeenth-century Europe both as a normative ideal (and representational space) about how politics should be organized geographically and as an alternative mode of socio-economic organization to imperial or 'node and network' (trading system) ideals. It was a central feature of the uniquely European development of an expansive capitalism that slowly emerged between the fifteenth and eighteenth centuries (Palen 1992). The development was fastest where city-states were least well-established on the seaward and continental margins of Europe (Rokkan and Urwin 1983). There was not a movement from one to the other, as argued by Mumford (1961), among others. On the contrary, territorial states became established most strongly where city-state development was least entrenched. Only in the nineteenth century were the last of the city-state territories converted into the territorial states of Germany and Italy.

The system of territorial states developed important legal and economic underpinnings as a wide range of spatial practices became more and more bounded by state-territorial limits. Through a social process of recognizing other spaces as potentially 'developed' and 'modern' insofar as they acquired the trappings of territorial statehood (armies, judiciaries, etc.), the state-territorial form of spatial organization came to encompass in some degree most of the world's population. Since the early nineteenth century and, more especially since the Second World War, international organizations, especially the United Nations, have played a fundamental role in formalizing this process (Luke 1993; Murphy 1994). From one point of view, a territorial state can be said to exist because its flag flies outside the UN headquarters in New York, irrespective of its effectiveness as a political–economic entity.

The actual spatial organization of the world, however, has always been more complex than the simple assimilation of all social cleavages into a superordinate state-territorial spatial form assumed by most varieties of thinking about international political economy (Chapter 4 is devoted to this topic). In particular, the spatial practices of everyday life have maintained a place-specificity that defies the intellectual assumption of growing state-territorial homogeneity. The recent explosion of regional–ethnic politics around the world bears witness to this. More generally, social groups are often defined by their spatial configurations: their relative spatial isolation and claims to territory are the root and symbol of their existence. The persisting segregation of American ethnic and racial groups is one example of this. Furthermore, the globalization of the world economy in recent years has challenged the state territory as the basic building block of the world economy. The financial and informational flows at the heart of the new economy are not readily contained by territorial-state boundaries. They favour or disadvantage particular regions and localities within states rather than national territories in their entirety (Agnew 1987b).

How can we best begin to make sense out of the historical evolution of the geographical basis to the international political economy? At the risk of dramatically oversimplifying a much more complex reality, we identify three geopolitical orders over the period 1815–1990 in which different mixes of spatial organization have been predominant. This is an interpretative act and open to challenge, but we find it a helpful way of placing contemporary trends into a historical framework.

THE CONCEPT OF GEOPOLITICAL ORDER

In our usage 'order' refers to the routinized rules, institutions, activities and strategies through which the international political economy operates in different historical periods. The qualifying term 'geopolitical' draws attention to the geographical elements of a world order. This is not some 'special' or 'extra' feature of an abstract order. Rather it is intrinsic to it. Orders necessarily have geographical characteristics. These include the relative degree of centrality of state territoriality to social and economic activities, the nature of the hierarchy of states (dominated by one or a number of states, the degree of state equality), the spatial scope of the activities of different states and other actors such as international organizations and businesses, the spatial con-nectedness or disconnectedness between various actors, the conditioning effects of informational and military technologies upon spatial interaction, and the ranking of world regions and particular states by the dominant states in terms of 'threats' to their military and economic 'security'.

From this point of view 'order' does not imply a world based on consensus or cooperation. A normative conception of order, as in an 'ordered world without conflict', can inspire criticism of existing arrangements. But there is a clear difference in meaning between describing the sets of arrangements upon which *empirical* order rests and judging the quality of those arrangements. This distinction is important. At the present time there is controversy over the possibility of a 'new world order' and whether this will lead to a 'better world'. In Chapter 8 we deal with this normative conception of order directly in a discussion of our favoured scenario for future world order.

Here our interest is analytic. Any global geopolitical order is a mixture of cohesion and conflict between actors; what is required for a global order to exist is some organized system of governance: definition of actors, rules of operation, principles of interaction, and widely shared assumptions about trade, force and diplomacy. These would include formal international organizations and 'regimes' covering the behaviour of governments in specific issue areas (such as trade, money, security, etc.). But a world order also requires a set of intersubjective assumptions and behavioural orientations shared by leading actors. In this usage disorder exists when a system of global governance (however conflictual and including economic management) breaks down and the change that undermined the old order has not yet produced a new one.

A basic assumption here is that the identity and interests of states (and other actors) are formed in interaction with one another. A state cannot exist and act as such unless it is recognized by others. Wendt (1992, 396–7) provides the following argument for the social origins of statehood:

> A fundamental principle of constructivist social theory is that people act towards objects, including other actors, on the basis of the meanings that the objects have for them. States act differently toward enemies than they do toward friends because enemies are threatening and friends are not. Anarchy and the distribution of power are insufficient to tell us which is which. U.S. military power has a different significance for Canada than for Cuba, despite their similar 'structural' positions, just as British missiles have a different significance for the United States than do Soviet missiles. The distribution of power may always affect states' calculations, but how it does so depends on the intersubjective understandings and expectations, on the 'distribution of knowledge', that constitute their conceptions of self and other. If society 'forgets' what a university is, the powers and practices of professor and student cease to exist; if the United States and Soviet Union decide that they are no longer enemies, 'the cold war is over.' It is collective meanings that constitute the structures which organize our actions.

While agreeing with Wendt's insight about the role of intersubjectivity in creating statehood and the state system, the inspiration for our conception of periods of distinctive geopolitical order comes from the political economy of Cox (1981, 1987) which sees the activities of states and other actors as aspects of 'larger historical structures'. From this point of view, states are historically constructed (and reconstructed) in the nexus between global and domestic/

local social relations. State power is the capacity for action in historically specific structural contexts. Proceeding from this view, Cox introduces the concept of hegemony, as developed originally by the Sardinian–Italian communist thinker and leader Gramsci, to refer to the creation of distinctive structures or orders based upon shifts in the social organization of the world economy. These orders are maintained by a mix of power relations, both coercive and consensual. However, perhaps the most important feature of this perspective lies in its emphasis on the routinized and incorporated nature of the practices and ideological representations that give an order its 'normality' and 'commonsensical' acceptability to the actors involved in it.

This notion of hegemony or cultural–political order departs significantly from another usage of the term, when it is used simply to refer to the domination exercised by a single state over all other states or over a particular historical epoch. Particular states may well be the agents of hegemony, as will be argued later, but *there is no necessary requirement for a period of geopolitical order to be associated with domination by a single state*. At the very least, global hegemony presupposes the establishment of a dominant historic bloc of élites in different states that accept fundamental premises about the nature of the world economy and inter-state relations. But we would argue that in the Gramscian sense a period of geopolitical order cannot be regarded as 'non-hegemonic' simply because the hegemonic practices and ideas of the period are not identifiable with a single state dominant at the global scale. Indeed, there may well be 'competing' hegemonies (and aspiring hegemons) based in different states that, while confronting one another with different modes of socio-economic organization, tacitly accept the same assumptions about the nature of statehood and the rules of international behaviour. This was the case to a degree with the United States and the Soviet Union during the Cold War. As Gill (1990) has argued, hegemony can also be vested in the operations of international organizations. In emphasizing these cases we would differ with those, both 'Gramscian' and others, who reserve the term hegemony only for periods of single-state domination. This is an unnecessarily restrictive assumption which implies that without a single-state hegemony the world economy and inter-state relations revert to an anarchic 'state of nature' which has no common features based on shared rules, practices and understandings. In our usage, therefore, hegemony refers to a cultural complex of practices and representations associated with a particular geopolitical order without the requirement of a dominant territorial agent. There is always hegemony, but there are not always hegemons.

Yet order clearly implies the presence and operation of standardized or regularized rules of behaviour. Bull's (1977) criticism of the sharp distinction between hierarchy and anarchy as organizing principles for inter-state relations offers an interesting starting point for a discussion of the nature of order. He provides a contrast between what he terms a 'system' of states and a 'society' of states. In the former there is nothing more than interaction –

17

usually war – between the basic units. The Romans and the Vandal invaders of the Roman Empire would be examples of units in such a system. After 1815, however, a 'society' of territorial states based around the modern conventions of state recognition and diplomacy, balance of power, and war prosecuted to enforce 'community norms' came into full existence for the first time, even though it had older roots. The point here is not to deny the coercion and violence associated with the post-1815 society of states. It is only to suggest that there were now 'entrance requirements' for statehood not found in the older 'system' of states. The most important of these was the mutual recognition of legitimate command or sovereignty over a discrete territory (see also Giddens 1984; Thomson 1994).

Bull's emphasis on 1815 as a watershed is doubtless overstated and rather ignores the significance of the Treaty of Westphalia in 1648 for the establishment of the European state system. But it was only after the Napoleonic Wars that 'power politics' in the modern sense of the term – associated with the concept of a separate 'foreign policy', standing and conscript armies drawn from politically mobilized populations, and the balance of military power between territorial states as an operative principle – finally emerged as the defining feature of global political relations (Hinsley 1966, ch. 4). Bull also sees these new 'rules' governing state conduct only in the regulative sense of limiting or constraining an individual state within the society of states. The older 'system' of states was apparently ruleless. We prefer to see rules as performing a constitutive role (Dessler 1989). Even under apparent anarchy there are rules or conventions that provide a stock of standardized practices and representations for international communication and behaviour. The idea of action by states or other actors presupposes 'conventions of meaning' that are consciously or implicitly shared by all parties including the self-definitions of the parties themselves. The 'rules-of-the-game' provide the basis for the more substantive tacit agreements, informal but implicit bargains, and formal treaties that regulate international behaviour in particular historical epochs (Keal 1984; Kratochwil 1989; Onuf 1989; Lipson 1991).

Rather than the result of deliberate calculation by a central 'directing agency' or intelligence, geopolitical order arises out of the spontaneous actions of states and other actors. What arises is not necessarily what each intends. 'That the social world is a world of meanings, purposes and intentions does not always mean that acting on those meanings creates expected results' (Inayatullah 1993, 26). This potential discrepancy between intentions and outcomes helps to undermine a given geopolitical order over time. 'Transition' from one order to another, therefore, involves a transformation of the 'common sense' understandings of how international relations work and what is at stake for particular parties as 'the position-practice system' (or 'habitus' to use Bourdieu's (1977) term) linking intentions and outcomes changes over time. A geopolitical order, therefore, is always partial and precarious, achieved through social practice rather than imposed through a transhistorical logic.

Of fundamental importance, however, and missing from purely constructivist accounts of states and their interactions, are the technological and economic circumstances of different eras that define the geographical contours of opportunity and constraint which set limits to both human intentions and their achievement (Deudney 1993). Geopolitical orders, therefore, are not simply products of interaction among social actors. They rise and fall in relation to changing technological and economic conditions. There is a materiality to the hegemonic concepts which define the 'normalcy' of each geopolitical order (Overbeek and van der Pijl 1993). For example, the overall balance between financial business interests and industrial business interests and *the relative geographical scope of their operation* (local, national, world-regional, global) in the world economy helped create the conditions in which 'liberal internationalism' (1820–70), 'state monopolism' (1870–1945), 'corporate liberalism' (1945–70s) and 'transnational liberalism' (1970s–90s) became the hegemonic concepts of their respective eras.

The concept of geopolitical order, therefore, rests on three premises. The first is that since the emergence of the modern 'society' of territorial states in the early nineteenth century, the rules, practices and ideas governing the international political economy have changed historically and a periodization to these changes can be identified. The second is that these rules, practices and ideas are sociologically grounded in the interaction between states and other actors. As the nature of interaction changes so do the practices and ideas. They cannot be reduced to the transcendental drives or instincts offered by, for example, concepts of 'rational choice' or *'realpolitik'*. The 'interests' of actors arise out of historically-specific contexts of action. The third is that such changes in practices and ideas are intrinsically geographical; they involve shifts in the differentiation of the spatial fields of practice that are the root and symbol of a geopolitical order. Geopolitical order, therefore, does not refer to the apostolic succession of Great Powers to hegemonic status but to the changing geographical basis to the international political economy in different historical periods.

THE THREE GEOPOLITICAL ORDERS

In using the concept of geopolitical order it is important to identify the criteria and establish the time-limits used to define each order. The first period (1815–75) is one of a European territorial balance of power in which Britain came to command and define the nature of the growing world economy outside of Europe (the Concert of Europe – British Geopolitical Order). Britain held the balance of power in Europe, enjoyed a significant edge in seapower that allowed it a coercive role in imposing its policies around the world, and sponsored a set of economic doctrines – comparative advantage, free trade and the gold standard – that, while appearing universal, benefited influential interests in Britain.

This combination of European Concert and British domination outside Europe gave way in a second period (1875–1945) to a destabilization of the European balance of power as certain other states (especially Germany) challenged British policies. National economies became increasingly autarkic and protectionist and the world economy lost its emphasis on free trade and the gold standard and divided into economic blocs. The two World Wars were an intrinsic part of this drift towards a state-territorial international political economy (the Geopolitical Order of Inter-Imperial Rivalry).

In the third period (1945–90), a Cold War Geopolitical Order arose out of the ashes of the Second World War. The two principal victors in that war, the United States and the Soviet Union, divided the world into two spheres of influence with different political economic models of development (the First World of capitalist organization and the Second World of state-planned organization) and competed for influence in a third (the Third World of new states emerging from colonialism). American hegemony involved a structure similar to that sponsored by Britain in the nineteenth century but with much greater commitment to opening up the world to direct investment and trade, and with a security apparatus that was much more extensive and intrusive than that exercised by Britain in its day. This neo-liberalism was also much more institutionalized in a set of security alliances (especially NATO) and global trade, investment, and monetary arrangements (GATT, the IMF, the World Bank, etc.) than had been the British variety. The net effect of the American-based hegemony, however, has been to create a 'transnational liberal order' that is no longer uniquely associated with the United States. As this occurred, and with the disintegration of the Soviet Union and its sphere of influence in 1989–90, it signified the collapse of the Cold War Geopolitical Order. Only the outlines of a new geopolitical order can be glimpsed as yet, though it is clear that a new form of transnational liberalism will be central to it. (It also goes without saying that the beginnings and ends of each of our geopolitical orders embrace a period of 'morbid disorder', as in the 1940s and 1990s. Each geopolitical order emerged from the unravelling of the previous one without the sudden transition that our use of precise dates might be taken to imply.)

The criteria used to define each of the orders can be identified more formally. Most of the contributions to a discussion of geopolitical order have been very selective, taking a single criterion (e.g. technology, trade, security, or money) as fundamental. In contrast, the framework outlined here is multicausal, following the movement in much contemporary social science from privileging single transcendental factors to exploring causality in its historical manifestations (e.g. Knox and Agnew 1994; Lash and Urry 1994). An attempt to achieve this is sketched in Table 2.1. Much of the rest of this chapter is an expansion on what is outlined here paying special attention to

Table 2.1 A framework for the analysis of a geopolitical order

	Global level	*State level*
1 World economic structure	International division of labour (IDL) Technological paradigms (TP)	Position in IDL Development of TP
2 Political– economic regulation		Financial system Interfirm relations Wage/industrial relations
3 Political– institutional forms	*International regimes*: trade, money, security systems	*Political system*: legal, military, political mobilization, govt, education systems
4 Mechanisms of order-establishment and maintenance	Endogenous learning Emulation Legitimation, coercion	
5 Geographical scale of economic accumulation	Territorial Interactional	
6 Space of political regulation	National state Imperial state International state	

Source: After Mjoset 1990

how the various criteria come together in different mixes to create the conditions for the establishment and maintenance of the three geopolitical orders.

The first four rows in Table 2.1 refer to the structural–institutional factors that have contributed to the origins and maintenance of the three geopolitical orders we identify. Our main argument is that shifts in geopolitical order correlate with cyclical and quantitative changes in these factors in combination rather than with a single factor in isolation. The bottom two rows in the Table identify the geographical trends in scale of economic accumulation and space of political regulation that correlate with the various mixes of structural–institutional factors. The two columns represent the two primary geographical levels – the global and state levels – between which past geopolitical orders have been constructed.

World economic structure refers to both the international division of labour and technological paradigms. The former refers to the geographical distribution of different economic activities – manufacturing, resource production, etc. – between and within states. The latter refers to the emergence and capture of the benefits of the new technological paradigms that some commentators believe have appeared approximately every 50 years since the 1780s (Modelski 1987; Marshall 1987; see Beenstock 1984 for a critique).

Political–economic regulation describes the various arrangements concerning industrial relations, interfirm relations and financial regulation that have evolved within states to regulate economic transactions and which can act to encourage or retard adaptation to the world economic structure. To the extent that these arrangements are exclusivist, for example in discouraging inward or outward investment, then, all other things being equal, they would encourage a more territorialized international political economy.

Political–institutional forms refers, at the state level, to the main features of different political systems and their conceptions of social order and, at the global level, to the institutionalized regimes governing trade, monetary relations and military security that states can put into place. A world of states relies fundamentally on a perpetual struggle to define, impose and, if possible, legitimize rules of conduct through institutional channels. These are not usually universally accepted because states have different institutional histories. Some states and social movements have challenged the status quo. For example, Gaullist France, Khomeini's Iran and the movement of non-aligned states (led by India's Nehru, Egypt's Nasser and Yugoslavia's Tito) all challenged the Cold War Geopolitical Order in various ways. Each of the three geopolitical orders, however, has rested on a set of understandings shared by most states about the contemporary world order even when these involve military rivalry and economic protectionism.

States, or, more specifically, their political élites, come to participate in a geopolitical order in a variety of ways. The principal mechanisms of order establishment and maintenance range from endogenous learning (or independent acceptance of dominant norms), through emulation (or imitation) and legitimation (socialization), to coercion. For a specific state several of these mechanisms may be involved. A geopolitical order will usually be more stable in the long run if it relies mainly on legitimation, emulation, and endogenous learning and rather less on coercion. Geopolitical orders change as the historic blocs or social groups whose commitments are central to them are replaced or reformulated around new commitments and strategies.

A geopolitical order is organized as a result of the mix of the foregoing factors in terms of two geographical dimensions: the dominant scale of economic accumulation (or the geography of economic activity) and the dominant space of political regulation. The former has thus far had two tendencies: territorial (intensive) and interactional (extensive). The latter has had three tendencies: national state, imperial state, and international state. The territorial scale of accumulation is based around tightly integrated national economies. The interactional scale is more extensive geographically, involving world-regional and globalizing patterns of trade and investment. With respect to political regulation, the national state is the basic unit in a polyarchy or balance of power order whereas the imperial state can be said to exist when states seek to expand their control over discrete territories outside their current jurisdiction. The international state is one faced by the need to manage the growth of interactional accumulation.

The three geopolitical orders we identify have had specific combinations of these geographical tendencies. The first order (1815–75) rested on a set of territorial economies and national states in Europe with an expanding interactional tendency outside of Europe managed largely by Britain. From this viewpoint Britain was the first international state.

The second order (1875–1945) involved an explosion of inter-imperial rivalries among the leading European states, the United States and Japan for territorial control over the rest of the world economy. Initially, the competition between these imperial states produced a rapid growth in the world economy through the establishment of specialized colonial economies. But this was organized around exclusive zones tied to particular Great Powers. Eventually this gave rise to increased enmity from those states closed out of the enterprise (Germany, Japan and Italy). Failures of political–economic regulation at both state and global levels enhanced the primacy of the territorial scale of accumulation. The low level of US participation in world trade reduced the incentive for American replacement of Britain as the manager of a world economy beyond the imperial trading blocs.

The third order (1945–90) was a combination of two imperial states in global military and ideological competition (the United States and the Soviet Union) with American sponsorship of interactional accumulation, particularly among the allied states of the First World. The American agenda for the US as an international state centred around the drive to prevent the re-emergence of the imperial blocs that many influential Americans believed had produced both the Depression of the 1930s and the Second World War. This involved the simultaneous pursuit of three strategies. The first was the demilitarization and economic reorientation of Germany and Japan. The second was the containment of the Soviet Union and its state socialist model of economic development through regional military alliances (such as NATO). The third was the creation of a set of international institutions to project American practices and ideas about political–economic organization at a global level. These included the United Nations system, GATT, the IMF, the IBRD/World Bank, and a variety of international agreements such as the Bretton Woods Agreement of 1944 governing international monetary relations.

After the 1960s these strategies began to produce a globalized world economy in which many states became progressively internationalized. The costs of maintaining global military antagonism in the face of an increasingly open and competitive world economy and the failure of the Soviet model to incorporate successfully the latest technological paradigm (information and communication), worked with other forces (including social movements) to effectively undermine the Cold War order. The *Pax Americana* was designed to promote economic interactions between non-socialist national economies; its very success encouraged a degree of globalization that has dented the powers of almost all territorial states, including the USA.

TWO ALTERNATIVE SCHEMES

There are two other approaches to geopolitical order that are worth mentioning both for their own importance and for what they say about the particularity of the framework we have provided. The first is that of Robert Cox (1987). He too works with three geopolitical orders: 1845–75, 1875–1945 and 1945–65, but he also identifies an incipient order under construction since 1965. He argues that world orders and the roles of the state are grounded in social relations. His periodization likewise identifies the degree of geographical fragmentation or integration of the world economy as an important indicator of the character of a specific world order.

But Cox's dates are a little different: the first period does not begin until 1845 and the third period is over in 1965. These differences are not arbitrary. They reflect two significant differences between our approach and that of Cox. First, he sees the first and third periods as hegemonic in the dualistic sense of a dominant mode of social organization sponsored by a particular state: Britain in the first and the United States in the second. The second period and the present one are viewed as 'non-hegemonic'. This seems to imply, in the former case at least, that there is a 'reversion' to anarchy rather than, as we have suggested, a geopolitical order without a dominant territorial agent. Such time-periods as 1815–45 (in Europe) and 1875–1945 (everywhere) seem to fit uncomfortably into schemes that have a rigidly territorial conception of hegemony. Cox's understanding of hegemony involves a particular reading of Gramsci in which the state serves as the agent of hegemony. As Anderson (1976/77, Part I) points out, this is only one of three understandings of hegemony that can be derived from Gramsci's writings. A more pervasive one locates hegemony in the action of groups in (civil) society. In this usage, therefore, hegemony can exist without state agency (also see Augelli and Murphy 1993).

Second, the US–Soviet conflict figures as a backdrop to, rather than as a central feature in, Cox's account of the third period. His focus is upon the political economy of the *Pax Americana* and its breakdown in the 1960s. The Cold War rivalry which gave the *Pax* much of its ability to co-opt other powerful states is seen as dissipating more rapidly than we believe recent history gives grounds for. Further, the imperial rivalry between the United States and the former Soviet Union not only provided an important stimulus to the creation of the 'Free World' economy from which recent changes in the international political economy have emanated, but it also gave rise to a militarization and external orientation of many Third World states that has not yet ended. The other side to the economic 'globalization' of the 'core' industrial states has been the militarization of many regions in the 'periphery'.

A second approach is provided by Peter Taylor (1993a). His framework draws directly on Wallerstein's understanding of hegemonic cycles and Modelski's work on long cycles (see Tables 2.2 and 2.3). In this usage hegemony is centred on the activity of a dominant state and its ability to

Table 2.2 Wallerstein's cycles of hegemony

Hegemony	World war securing hegemony	Period of dominance	Decline
Dutch	Thirty Years' War, 1618–48	1620–50	1650–72
British	Napoleonic Wars, 1792–1815	1815–73	1873–96
American	World Wars I and II, 1914–45	1945–67	1967–

Source: Wallerstein 1984, 41–2

Table 2.3 Modelski's long-cycles of world leadership

Cycle	Global war	Preponderance	Decline
1495–1580	1494–1516	Portugal, 1516–40	1540–80
1580–1688	1580–1609	Netherlands, 1609–40	1640–88
1688–1792	1688–1713	Britain, 1714–40	1740–92
1792–1914	1792–1815	Britain, 1815–50	1850–1914
1914–	1914–45	United States, 1945–73	1973–

Source: Modelski 1987, 40, 42, 44, 102, 131, 147

combine economic and military superiority into command over the world economy. From this point of view a series of hegemonic powers has emerged from the sixteenth century onwards to give command and control to the international political economy. A recurrent rhythm of boom, crisis and restructuring has followed from dual processes of capital accumulation and hegemonic rivalry. Geopolitical order refers to the impact of a change in the identity of the hegemonic power (or a period of the transfer of hegemony) plus the geographical ordering of economic and military priorities by the hegemon (geopolitical codes) rather than to the changing substantive geography of the international political economy. In this important respect our approach is much closer to that of Cox with its emphasis on the multicausal origins of the shifting fragmentation and integration of the world economy than it is to that of Taylor. Ironically, given the avowed emphasis on states as entities embedded in the long cycles of the capitalist world economy, Taylor's approach endows 'key' states with an essential nature and an exclusive role in the formation of all geopolitical orders. Taylor has little to say about the impact of changing economic contexts on geopolitics.

The central theme in Taylor's scheme is the hegemonic succession to Britain. Two twentieth-century geopolitical orders are identified: the World Order of the British Succession (1907–44) and the Cold War World Order (1946–89). The first had a precursor in the World Order of Rivalry and Concert (1871–1904) which in turn had replaced the World Order of Hegemony and Concert (1815–66). In one sense this periodization is merely a more refined version of the one we have proposed, particularly in relation to the period 1875–1945. It certainly reflects a stronger attachment to the inflection points of the long-wave cycle; each phase must be isolated and named. But it also evidences a certain teleological insistence that a (eternal?) 'struggle for hegemony' (e.g. the 'British Succession') is what drives the creation of a geopolitical order rather than the technological–economic cycles the periodization relies upon. Contra Cox, this does at least allow Taylor to argue for a hegemony immanent within a non-hegemonic order!

The principal drawback of this approach, however, lies in seeing a geopolitical order entirely in terms of the coercive potential of actual and potential hegemonic powers rather than as a historical structure based on the prevalence of certain practices and representations. What it means to be a state or how a geopolitical order works are not given for all time and in abeyance of the workings of the international political economy. These are the primary distinguishing features of our approach when compared to that of Taylor.

THE CONCERT OF EUROPE – BRITISH GEOPOLITICAL ORDER (1815–75)

The British geopolitical order of 1815–75 provides strong evidence for the view that there are alternatives to rules of governance manifesting either total anarchy or strict hierarchy. The international political economy between 1815 and 1875 was characterized by a European Concert in which no one state 'laid down the law' for the continent as a whole and an emerging British economic hegemony in much of the rest of the world. Putting it another way: 'Great Britain's role as a world power did not translate into continental hegemony. The governance system in nineteenth century Europe was a polyarchy, not a hegemony' (Holsti 1992, 56). These twin features are what sets this period apart from later ones. After 1875 both the Concert and certain vital features of British economic hegemony decayed rapidly as the fundamental supports of the order collapsed.

The main problem the Concert was designed to handle was revolution. Initially there was agreement that this meant preventing a French war of revenge or the restoration of the regime of Napoleon Bonaparte. Thereafter there was disagreement over the extent to which this justified unilateral interventions in small states to suppress rebellion. But a consensus evolved among the dominant political élites in Europe that (1) no one state in Europe could predominate within the continent and (2) Europe-wide wars were best avoided because of their potential for unleashing revolutionary forces. Even though the system of regular congresses largely collapsed, certain rules of behaviour became widely accepted. Unilateral responses were decried. The Great Powers were co-managers who merited consultation on all of the principal threats to international peace. As a result, wars to foment revolution or to acquire territory were seen as threats to a system based on a conception of a collective interest that transcended particular national interests. In practice many conflicts which would have found their way easily to the battlefield in the eighteenth century were prevented by concerted diplomacy.

Britain not only experienced an Industrial Revolution before the rest of Europe, it also traded with and invested in other continents on a much larger scale than other European countries (save possibly for the Netherlands). These developments were connected. The technical innovation and organizational efficiency of early nineteenth-century British industry produced goods for export and capital for investment overseas and a demand for certain raw materials (especially cotton). In addition, however, Britain acquired a competitive advantage over other European states in the growing world economy. Dehio (1962, 71) sees this as arising out of a conjuncture of factors equivalent to those which established Venice as the previous great mediating agent within the world economy:

> Both in overseas trade and in naval strength, Britain gained supremacy, favored, like Venice, by two interacting factors: her island position and the new role which fell into her hands, the role of intermediary between two worlds. Unlike the continental powers, Britain could direct her undivided strength toward the sea; unlike her Dutch competitors, she did not have to man a land front.

The British orientation towards external expansion contrasted throughout this period with the Euro-centredness of the other European Great Powers. They were much too caught up with national economy-making, a process largely completed by Britain, to challenge British pre-eminence in the making of the world economy outside of Europe. By progressively opening up their domestic market from 1823 to 1840 through the sponsorship and unilateral practice of freer trade, British leaders created a world-wide network of trade and financial flows that presented their centrality as a by-product of the workings of a 'world market' that operated to the benefit of all. British governments could claim with credibility that the growth of the British economy served not only a national interest but a 'global' interest as well. This argument appealed particularly strongly to those upwardly-mobile social elements in settler (and former) settler colonies and social groups elsewhere involved in trading and financial activities.

Under British auspices market exchange was effectively globalized as production for the market replaced the mere trading of goods. The British national economy became the locomotive of the world economy. As industrialization proceeded in Europe and markets there became more competitive, British capitalists were pushed into widening their markets elsewhere. The British Empire was an important part of this 'internationalization', but the United States, China and Latin America, regions outside formal British control, were also important (Table 2.4). The internationalization of the British economy in the nineteenth century was a crucial element in the quickening pace and increasing spatial scope of the world economy.

Table 2.4 The geographical development of the world economy in the nineteenth century

Stage: Factor intensity:	Developed	Developing		Undeveloped	
	Capital	Labour	Land	Land	Labour
1800	Britain	Europe		USA	India
1840	Britain	Europe	USA	Latin America Australia Canada	India China
1870	Britain, Europe		USA	Australia Latin America Canada Africa	China
1900	Britain Europe USA		Australia Canada Argentina Mexico S. Africa	Latin America Africa	India China

Source: Hansson 1951, 59–82

The recycling of capital returned from world-wide investments made London the premier world financial centre. Finance capital's main political interest lay in discouraging economic disruption and social revolution in those places where it had investments. British governments not surprisingly came to value 'stability' and frequently intervened militarily through 'gunboat diplomacy' to restore or extend the status quo. British 'free trade–imperialism', as this mix of force, trade, investment, diplomacy and ideology has since been called, represented the emergence of a new kind of state that expedited the growth of international economic transactions.

An interesting case study in the effectiveness of British leadership in enlarging the world economy is provided by the passage of the Walker Tariff in the United States in 1846. This legislation reversed a previous commitment to protectionism by opening American markets to British manufactured goods. It was the product of a coalition of Southern planters and Western grain farmers interested in promoting export agriculture. Although this coalition was not to last, its existence shows the possibilities for attracting foreign allies that the British were not shy at exploiting. By repealing its Corn Laws in 1846 and allowing access to its grain market, Britain enticed the Western farmers of the United States into a political coalition they had hitherto avoided. The net effect was to encourage freer trade through the use of international 'market power' which altered the balance of economic interests and political power within another state. So, rather than just coercion or ideological hegemony the British also exercised their leadership through changing the structure of opportunities for international as opposed to domestic sales (James and Lake 1989).

There is some danger in overstating the achievement of freer trade during the period 1815–75. There were none of the multilateral agreements governing international trade that became institutionalized after the Second World War. Until the 1840s Britain itself remained largely protectionist and derived a considerable share of government revenues from customs duties. The era of freer trade only started in earnest in 1860 when Britain and France signed the Cobden–Chevalier Commercial Treaty. This treaty was by itself a bilateral agreement that allowed each party to retain a variety of tariffs and restrictions against other countries. However, it did contain a most-favoured-nation clause that required Britain and France to give each other any benefits they received from agreements with third parties. This did stimulate a degree of multilateral trade liberalization. But the spatial scope of the treaty proved restricted. Many of the world's leading economies failed to join the bandwagon, among them the United States, and the 'free international' trading system never extended beyond a limited number of the leading national economies (Stein 1984).

Evidence for a more obviously successful (and coercive) approach to encouraging open trading for all can be seen in British policies towards Turkey, China and its own territorial empire. Under an 1838 commercial treaty Turkey was required to allow British ships the right to sail through the Bosporus from the Mediterranean into the Black Sea, abolish internal duties on trade within the Ottoman Empire, and grant British exports most-favoured-nation status. In China British gunboat diplomacy successfully opened the country to international trade, initially as a result of British intervention to protect the trade in opium from India. The Treaty of Nanking (1842) led to Chinese diplomatic recognition of Britain, the British acquisition of Hong Kong, the opening of five ports to British traders, and the grant of most favoured nation trading status to British exports. Within the British Empire after 1849 trade was open to traders from all states, thus providing an example of consistency in the application of the doctrine of free trade that was widely commended by British advocates (O'Brien and Pigman 1992).

Britain's frequent use of coercion outside of Europe (but not in relation to the United States) and limited success in sponsorship of free trade among the other European Great Powers should not be allowed to obscure the fact that in the 'liberal interlude' from 1846 to 1875 Britain managed to turn free trade into the dominant philosophy of international economic management. Moreover, a large expansion in world trade and an economic boom in Britain coincided with the movement towards trade liberalization. It was not difficult in Britain, and elsewhere, to see a causal connection between free trade ideology and the explosion of growth. Ironically, under conditions of economic downturn from 1873 to 1896 the British commitment to unilateral free trade after 1870 probably helped undermine Britain's own economy as other national economies failed to follow suit and, indeed, embarked on neo-mercantilist policies that protected their economies at the expense of the British. However, it was not until 1914 that the liberal trade regime established in the mid-nineteenth century was definitively abandoned. Indeed, world trade expanded at unprecedented rates in the fourteen years after 1896 (Stein 1984).

In the 1870s Britain began to lose its central position within the Concert of Europe and also to sacrifice its industrial strength on the altar of free trade. In both respects, the political–economic rise of Germany and the explosive economic growth of the United States were the most significant developments. British concentration on the 'old' industrial technologies of the steam era and its increased commitment to such geographically extensive activities as commerce and finance put it at a disadvantage relative to the two massive territorial economies with which it now had to compete. Neither country had Britain's fervent commitment to free trade. The newly unified Germany, in particular, was not willing to accept political subsidiarity to Britain, either in Europe or outside.

Fatefully, by 1875 Britain had also turned away from Europe and the United States towards its empire and those world regions where its hegemony was apparently more secure. This provided the impetus for the collapse of the Concert in Europe and the beginning of a British retreat from its constitutive role in the international political economy. Britain certainly maintained its absolute centrality to world commerce and finance for many years but it was without the almost unquestioned position of supremacy it had established during the period from 1815 to 1875.

One of the most important aspects of the Concert constructed in 1814–15 was its territorial equilibrium of power (Holsti 1992, 53). The main measure of state power was maintenance of national territory. Differences in terms of military might did not excite much concern and did not threaten the balance of power. Throughout the period Russia had by far the largest army and Britain had a substantial naval supremacy. These constants imposed a stability that until the 1860s made the Concert a territorial rather than a military balance of power. Thereafter changes in military technology evened out military potential and led to an emphasis on mobility, mobilization and preparation for pre-emptive attack. Relative military strength

> became the measure of the balance of power, and any increment was perceived as threatening to the equilibrium. Arms racing gave the appearance of rapid changes in relative power and generated fears of lagging behind. . . . Industrial dynamics replaced territory as the main metric of power analysis.
>
> (Holsti 1992, 53)

A parallel change took place in conceptions of warfare. In the aftermath of the Napoleonic Wars engaging in war required a defined political purpose. Clausewitz was the chief proponent of this view (Pick 1993). But by the 1870s armed conflict was no longer seen as a final step in a deliberate escalation from intensive diplomatic and conciliatory activities. War became an unavoidable and purposeful activity intimately related to an 'eternal' struggle between nations and social groups. Intellectual authorities such as Hegel and Darwin were drawn from (and distorted) in justifying warfare as encouraging patriotism, social cohesion and national purpose (Holsti 1992). Of particular importance, organized warfare was increasingly seen as an antidote to the 'degeneration' many intellectuals saw as intrinsic to modern, industrial society. Anarchy and anomie were the ever-present threats to the progress which proponents of evolutionary naturalism saw as locked in struggle with degeneration (Pick 1989). In its capacity for defining a national purpose and generating a spirit of sacrifice war was therefore regenerative and redemptive.

Perhaps of greatest importance, however, the Concert of Europe, though based initially on a set of states that had been active in European international relations for many years, eventually saw the addition of many new states that had originated in revolts against established ones. As the century wore on the Concert had to come to terms with the creation of new states from the fission and the fusion of its original members. Beginning with the emergence of Greece from within the Ottoman Empire in the 1820s, through the unification of Italy (at the expense of Austria) and Germany (at the expense of Austria and France), to the explosion of the Balkans (against the Ottoman and Austrian Empires) at the century's end, the creation of the new states effectively called into question the 'principles' upon which the Concert had been based. The new states had no commitment to the old order. In their wars of independence and frequent attempts at territorial expansion they effectively undermined it.

The order had been designed to tame and repress the nationalism that spread throughout Europe after the French Revolution. Its founding states were territorial states. But they were organized around aristocratic and bourgeois élites rather than mobilized national groups. Throughout the nineteenth century a new conception of statehood based on the creation of nation-states came increasingly to the fore. These were states built upon ethnic/religious/language divisions and particularities. In the identity of state with nation, territorial sovereignty became fused with the fate of the nation. The 'interests' of peoples were rigidly territorialized as the number of states in Europe tripled. This new system of states had no place for the rules that had governed the previous one.

THE GEOPOLITICAL ORDER OF INTER-IMPERIAL RIVALRY (1875–1945)

The German and American national economies worked very differently from the British national economy and from one another. What they shared was a competitive advantage in the size and dynamism of their domestic markets. The British economy was much more dependent on international transactions and increasingly lagged behind them in the new 'cutting edge' electrical and chemical industries. But they differed in that the American economy had become the main beneficiary of British free-trade imperialism, joining its now continental dimensions and vast resource base with access to British mediated capital flows and credit. Germany's economic development was based upon a state-sponsored industrialization that gave priority to military goods. Its external financial links remained heavily orientated to Britain.

This was an inherently unstable situation. Germany's military–industrial capability could not be translated into an enhanced global political role without upsetting both the Concert and the global flows of trade and capital centred around Britain. At the same time the increased industrial production of the United States and the European states undermined British industrial pre-eminence and led British business and governments into the use of political dominance and non-tariff barriers within its empire to restrict price competition and monopolize trade. The net result was an erosion of the system of interactional accumulation based around a set of national states and one international state (Britain) that had predominated previously and the emergence of a set of competitive imperial states dividing the world into zones based on territorial accumulation.

The inter-state system quickly polarized into two antagonistic groups. One, headed by Britain and France (with tacit American support), was orientated towards maintaining free-trade imperialism. The other, headed by Germany, was concerned with expanding its territorial possessions and challenging British financial hegemony. This division was apparent by the 1890s and gained its most famous early expression in the arms race between Britain and Germany known quaintly as 'the Anglo-German naval rivalry'. This polarization became the central feature of the geopolitical order as the ground for conflict shifted towards the territorial definition introduced by the Germans. It is no coincidence, therefore, that 1875–1945 produced the main peaks in, and the largest average annual number of, territorial changes, excluding those involving national independence, for the entire period 1815–1980 (Goertz and Diehl 1992, 41–4).

It is often contended that Britain fell from its zenith because when challenged by the Germans (and other potential enemies) it could no longer finance the military obligations that its empire and naval protection of free trade entailed. Available statistical evidence suggests that this was not the case. In terms of both percentage of national income and per capita spending for the years 1870–1910, and except for the years of the South African War, Britain had consistently lower figures than France and ones not much higher than Germany's. In addition, the 'burden' imposed by the empire was one Britain managed successfully to impose upon its colonies. The various parts of the far-flung empire paid for more than their own local defence. India in particular paid a substantial amount of Home Charges for the defence of the 'homeland'. As Offer (1993, 232) concludes his thorough survey of the evidence: 'overseas investment paid for its [India's] defence and left a substantial margin of profit.'

More to the point than the simplistic (and incorrect) 'imperial overstretch thesis' (Kennedy 1987) is the fact that the increasing economic importance of the British Empire to the British economy provided a powerful incentive to defend it against erosion from foreign competition. In return, the empire offered a reservoir of resources and labour for the prosecution of war that increased the capacity of the 'small island nation' to wage war against its challengers. The path to the two World Wars ran through London and Calcutta/Delhi as well as through Berlin.

Inter-state rivalry was powerfully fuelled by the growing nationalism of the period. National economy-making through the extension of railways and the reorganization of economic space into national blocs brought a collective sense of national economies as 'communities of fate'. The spread of literacy in standardized vernacular languages through universal elementary education enabled the more effective dissemination to mass publics of nationalized histories and myths of national origins (and superiority). State boundaries were seen to define natural units whose limits were the product of differences in national 'vitality' and 'capability'.

This vision was initially challenged by the growing socialist movements of the period. Class rather than nation was their basic conceptual category. However, they also succumbed to a pervasive nationalism by organizing nationally and having to adjust to the distinctive institutional arrangements governing industrial relations and political parties in different countries. Both ideologies (e.g. revolutionary versus evolutionary) and strategies (e.g. parliamentary versus industrial) became increasingly different in different countries. Perhaps most disabling was the lack of a 'global' vision. Appealing largely to the immediate material interests of industrial workers in the industrialized world, they were unable to articulate how subjugation and exploitation were now global in form rather than just factory-level or specifically national phenomena (though see Lenin 1971).

The late nineteenth century saw a vast enlargement and deepening of the world economy largely through the 'new imperialism' of the period. Whole regions inside and outside of Europe became specialized in the production of specific manufactured goods, raw materials and food products. The new spatial division of labour saw regional industrial specialization within the national economies of Europe and the United States paralleled by regional specialization in raw material production elsewhere. The downturn in the world economy in the period 1883–96, due among other things to decreased profitability in manufacturing, ushered in a spurt in the expansion of investment in raw material production. By 1900 not only was most of the world formally bound into the colonial empires which expanded rapidly after 1880 but more and more of the world's resources and population were drawn into a geographically specialized world economy. Bananas in Central America, tea in Ceylon (Sri Lanka) and rubber in Malaya are just three of the host of examples involved in this process of regional commodity specialization (Wolf 1982).

The outcome of the extension of the world economy and the intensification of trade within the regions already incorporated was a substantial increase in world trade. The contribution of the 'periphery' of resource economies to this growth was important. By 1913 Asia and Africa provided more exports than either Britain and Ireland or the United States and Canada. In 1913 Asia also had a share of world imports almost as large as the US and Canada combined (Table 2.5).

Table 2.5 World trade, 1876–1937
A World exports by geographical region, percentage of total
B World imports by geographical region, percentage of total

A

	1876–80	1896–1900	1913	1928	1937
USA and Canada	11.7	14.5	14.8	19.8	17.1
UK and Ireland	16.3	14.2	13.1	11.5	10.6
N.W. Europe	31.9	34.4	33.4	25.1	25.8
Other Europe	16.0	15.2	12.4	11.4	10.6
Oceania			2.5 ⎫	2.9	3.5
Latin America	} 24.1	} 21.7	8.3 ⎬ 26.3	9.8	10.2
Africa			3.7 ⎪	4.0	5.3
Asia			11.8 ⎭	15.5	16.9

B

	1876–80	1896–1900	1913	1928	1937
USA and Canada	7.4	8.9	11.5	15.2	13.9
UK and Ireland	22.5	20.5	15.2	15.8	17.8
N.W. Europe	31.9	36.5	36.5	27.9	27.8
Other Europe	11.9	11.0	13.4	12.5	10.2
Oceania			2.4 ⎫	2.6	2.8
Latin America	} 26.3	} 23.0	7.0 ⎬ 23.4	7.6	7.2
Africa			3.6 ⎪	4.6	6.2
Asia			10.4 ⎭	13.8	14.1

Source: Yates 1959, Tables 6 and 7, 32–3

With this global extension the international political economy acquired a new structure. The industrializing countries of Europe and North America bought increasing amounts of raw materials and foodstuffs from the economies of the 'periphery' and ran up large trade deficits with these regions. Britain, however, as a result of its free-trade policy, ran up substantial deficits as it imported manufactured goods from and invested heavily in the industrializing countries (especially the United States and Germany). In turn Britain financed its deficits through the export of manufactured goods to the so-called underdeveloped world, particularly its empire, Latin America and China. Thus the circle of international trade and finance was closed.

India and China were vital to this world pattern of 'cross-over' trade and payments. It was largely Britain's trade with India and China that compensated for its payments deficits with the United States, industrial Europe and the settler economies of Canada, Australia, South Africa and New Zealand. Without the 'Asian surplus' Britain would not have been able to underwrite the growth of these other economies. So, far from being peripheral to the growth of the world economy, the underdeveloped world, especially India and China, was central to it (Latham 1978).

The main axis of capital accumulation, therefore, was not within the 'core' of industrialized countries but rather in the core–periphery structure of the British Empire and the US–Britain relationship. The challenge for Germany was to build an alternative to this structure. Initially this was done by creating a colonial empire and using government support to help the spread of German private investment. Britain had long kept public diplomacy and private investment as rigidly separate activities. 'Laissez faire at home reinforced the separation of commercial and diplomatic activities abroad' (Lipson 1989, 193). But the active role taken by German and other governments in channelling investment by their nationals led British governments into a more aggressive role by the 1890s. The increased competition undermined free-trade imperialism as both trade and investment came to have direct government support.

One seeming advantage that Britain continued to enjoy was its centrality to the world financial system of the period. Indeed, the cross-over trading system required this. The large private banks and all the central banks of the leading industrial countries kept pound sterling accounts in London. As a result, all parts of the world economy were linked through fluctuations in interest rates established in London. Between 1890 and 1910 a gold standard became generalized across the leading national economies. Britain alone had full convertibility, and while this did express a certain strength it also indicated the relative weakness of the British position. The lack of a central bank in the US until 1913 led to the use of the London money market as a substitute. This introduced an important instability into the monetary system because of the seasonality of US demand for credit and the stock market panics and investment manias endemic to the US economy of the time. Increased security concerns also led countries to hoard gold. Taken together, increasing financial instability and competition for gold made it increasingly difficult for Britain (in the form of the Treasury and the Bank of England) to control the system it had called into being (Mjoset 1990).

Perhaps of greatest importance, however, the real economic centrality of Britain had been eroding since the 1870s. In this context the gold standard offered little if any means to protect it through seigniorage within the world financial system. Britain could try to export its economic problems (as it did to India especially: Keynes 1971), but it was more likely to import those of others. The monetary system weakened rather than gave added leverage to Britain within the world economy of the time.

By a strict definition the gold standard was not a 'regime'. It was not constituted by governments. Central banks and private bankers were its driving forces. From this point of view, 'high finance' coordinated the monetary regime and created a 'community of interest' in international security. In Polanyi's words (1944, 10), the inter-state network of high finance was the 'main link between the political and economic organization of the world' providing 'the instruments for the international peace system'. This appears contrary to the well-known view of Hobson (1904), who saw high finance as the villain of the day – creating a lack of investment in British production of goods and boosting the export of capital abroad where it was protected by different states which were increasingly in a growing competition with one another. In fact high finance was both dependent on states for protection of portfolio investment *and* had a stake in avoiding war. In 1914 it became obvious that the dependence on states had taken the upper hand.

Between 1914 and 1939 the system of multilateral trade and finance suffered a number of blows that progressively weakened it. One was the militarization of the leading European economies during the First World War which undermined trade within Europe. The second was the overall decline in trade between Britain and Asia. Successful industrialization in India and China displaced some British products (in particular iron and steel and cotton textiles: Tomlinson 1993) and Japanese competition undermined Britain's previously captive markets throughout Asia. A third problem was an overproduction of export crops and raw materials. Commodity prices fell and with them went a local demand for manufactured goods (Baker 1981).

The crisis of the cross-over trading system (along with credit policies in the USA) contributed to the spiralling downturn in the world economy that produced the Great Depression of the 1930s. The leading industrial states reacted to the Depression by raising tariffs and devaluing their currencies. These shifts in economic policy were premised on the assumption that Britain would remain 'open' as the linchpin of the world economy. However, as Stein (1984, 375) puts it: 'Depression left Britain unable and unwilling to accept an increasingly asymmetric bargain.' The imperialist regime of international trade established in the late nineteenth century was over. It took the Second World War to bury it completely.

The First World War can be seen as the result of the German challenge to British hegemony. Perhaps it could have been avoided or the underlying conflict managed in some other way if militarist attitudes (particularly commitment to the 'cult of the offensive') and the pervasive nationalism of the period had not been so strong. The Second World War can be seen in some respects as a repeat of the First. The treaties emanating from the First solved few of the tensions that underlay it. Rather, they added new sources of enmity and hostility between the principal European powers, especially in the form of the popular German view that Germany was excessively punished by economic reparations and territorial losses for its role in starting the war

(Keynes 1972). Further, the new states that were added to the system in Eastern Europe introduced not only a much greater potential for bilateral alliances (e.g. France and Poland, Britain and Czechoslovakia) but also a set of 'enemies' for Germany in a world region in which it had significant economic interests.

But what was most novel was that the 'club' of Great Powers was no longer predominantly European. It was the entry of the US in the First World War that proved decisive militarily. Japan was a leading power in Asia. On the eastern flank of Europe the revolutionary regime that established itself after the Russian Revolution of 1917 created a new kind of political economy that was claimed to represent the future for workers all over the world. None of these states could be effectively integrated into the decaying cross-over trading system with Britain as its linchpin or was able to help bring about a new system. In the 1930s Japan embarked on a territorial strategy of empire-building in East Asia. The United States, after sponsoring the collective security system called the League of Nations, withdrew from its active implementation. The Soviet Union saw itself as threatening to and threatened by the capitalist world economy. The 'system' was the enemy to overthrow and not something to be reformed or revitalized.

The climax came after the remilitarization of the German economy under Nazi rule. The fact that the Second World War involved the active alliance of Germany, Italy and Japan, the three Great Powers with the most autarkic economies and the ones whose élites were most dissatisfied with the global status quo, indicates the extent to which inter-imperial rivalry or competition for control over territory had replaced the previous combination of geographical modes of economic accumulation and political regulation.

THE COLD WAR GEOPOLITICAL ORDER (1945–90)

The completeness of the American-Anglo-Soviet victory over Nazi Germany and Imperial Japan had two immediate consequences. First, Soviet influence extended over Eastern Europe and into Germany. When the war ended Soviet armies were as far west as the River Elbe. This encouraged both a continuing American military presence in Europe and a direct confrontation with the Soviet Union as a military competitor and sponsor of an alternative image of world order. This was quickly to find its clearest expression in the geopolitical doctrine of 'containment', whereby through alliances and military presence the United States government committed itself to maintaining the political status quo established in 1945. The American development of nuclear weapons and a demonstrated willingness to use them meant that the security of the United States itself was beyond doubt (Art 1991). Indeed, the relative geographical isolation of the United States from most of its historic adversaries has always been an American advantage; if one discounts threats from nuclear armed terrorists or states that reject the 'norms' of inter-state behaviour (Table 2.6). What was in doubt in 1945–7 was the allegiance of other countries to the United States and its political-economic model.

Second, in economic and political terms the United States was without any serious competition in imposing its vision of world order on both its vanquished foes and most of its recent allies. Unlike after the First World War, when the United States turned its back on hegemony, this time there seemed to be no alternative. Europe and Japan were devastated. The reading of the origins of the Great Depression and the Second World War that predominated in the Roosevelt and Truman administrations suggested that the continued health of the American economy and the stability of its internal politics depended upon increasing rather than decreasing international trade and investment (Wachtel 1987). Europe and Japan had to be restored economically, both to deny them to the Soviet Union and to further American prosperity. Morgenthau's early-1940s plan for the 'ruralization' of Germany was quickly scrapped in 1945.

This is not to say that there was no opposition to the 'internationalist' position. Indeed, the Republican majorities in the US Congress in the immediate post-war years were generally as sceptical of the projection of the US New Deal experience of government economic intervention overseas as they were of its application at home. US forces demobilized rapidly after 1945. Only after 1947, with the growing fear of the Soviet Union as both foreign enemy and domestic subversive, did an internationalist consensus begin to emerge.

The period from 1945 to 1990 was one in which this consensus played itself out. The United States government set out in 1945–7 to sponsor a liberal international order in which its military expenditures would provide a protective apparatus for increased trade (and, less so, investment) across international boundaries. These would, in turn, redound to domestic Amer-

Table 2.6 Threats to US security in three eras according to Art (1991)

Type of threat	The Geopolitical era (pre-1945)	The Cold War era (1945–90)	The post-Cold War era (post-1990)
1 Invasion and conquest	Quite difficult after 1900	Practically zero probability	Practically zero probability
2 Slow strangulation through a global blockade	Of indeterminate feasibility	Practically zero probability	Practically zero probability
3 Nuclear attack			
From the Soviet Union		Not probable	Highly improbable
From other nations or subnational groups		Highly improbable	Not probable

ican advantage. The logic behind this lay in the presumed transcendental identity between the American and world economies. The expansion of one was seen as good for the other. Achieving this involved projecting at a global scale those institutions and practices that had already developed in the United States, such as mass production/consumption as a form of industrial organization; electoral democracy; limited state welfare policies; and government economic policies directed towards stimulating private economic activities (Maier 1978; Rupert 1990). Ruggie (1983) calls the normative content of these policies taken together 'embedded liberalism' because they were institutionalized in such entities as the IMF, the World Bank, GATT, and the Bretton Woods Agreement.

Three features of the American economy were particularly important in underpinning the internationalism of American policy. The first was economic concentration. Continuing an intermittent trend from the 1880s, in almost every American industry control over the market came to be exercised by ever fewer firms. Expanding concentration was accompanied and encouraged by the growth of government, especially at the federal level. Much of this was related to military expenditures designed to meet the long-term threat from the Soviet Union. These trends were reinforced by what became the main challenge to the perpetuation of the model within the United States: the direct investment of US corporations overseas. Much of this was in other industrialized countries. The axis of capital accumulation now ran through the core rather than between core and periphery. In the short run the repatriated profits benefited the American economy. But by the late 1960s, as domestic technology and management followed capital abroad, traditional exports were replaced by the foreign production of US affiliates to the detriment of employment in the United States. American mass consumption was no longer fully supported by the relatively high wages of its workers in mass production. This has come to define the crisis or impasse facing the American model in the United States (Agnew 1987b). What Arrighi (1990, 403) calls a Free Enterprise System – 'free, that is, from . . . vassalage to state power' – has come into existence to challenge the inter-state system as the singular locus of power in the international political economy.

The spread, acceptance and institutionalization of the American model was by no means a preordained or easy process. The key institutions and practices spread rapidly in the late 1940s and early 1950s. They were eventually accepted in all of the leading industrialized countries either through processes of 'external inducement' (Marshall Aid would be the classic example) and coercion (the British loan of 1946), or through direct intervention and reconstruction as in West Germany and Japan. In all cases, however, there was considerable compromise with local élites over the relative balance of growth and welfare elements in public policy (Ikenberry and Kupchan 1990).

Most of the key elements persisted until 1990, although some faded by the 1960s. They have been (1) stimulating economic growth indirectly through fiscal and monetary policies; (2) commitment to a unitary global market based on producing the greatest volume of goods most cheaply for sale in the widest possible market by means of a global division of labour; (3) accepting the United States as the home of the world's principal reserve currency and monetary overseer of the world economy (the Bretton Woods system, 1944–71; dollar-based floating exchange rate system, 1972–); (4) unremitting hostility to 'communism' or any political–economic ideology that could be associated with the Soviet Union; and (5) the assumption of the burden of intervening militarily whenever changes in government or insurgencies could be construed as threatening to the political status quo established in 1945 (the Truman doctrine).

Not only international relations, therefore, but also the domestic social order of other states was at issue in constituting the geopolitical order of the Cold War period. All states ideally were international ones; open to the free flow of investment and trade (Wood 1986). However, the system had a built-in flexibility that allowed both for national particularities and adjustments over time (Gilpin 1987). For example, and with respect to the former, the GATT is a trade regime initially formed in 1947 to encourage multilateral free trade, but it allows for domestic costs of adjustment. 'The GATT tries less to override . . . domestic political calculations than to moderate their impact on the larger network of liberal trade' (Lipson 1982, 425). The most-favoured-nation clause is a critical element in the GATT, as it was in mid-nineteenth-century trade agreements, 'but unlike its predecessor it is built on multilateral commitments, organized around regular consultations and periodic negoti-ations' (Lipson 1982, 426). Countries can escape from previous agreements if they can show that this will help their economic development.

It is little exaggeration to claim that in the five decades after 1945 American dominion was at the centre of a remarkable explosion in 'interactional' capitalism. Based initially on the expansion of mass consumption within the most industrialized countries, it later involved the reorganization of the world economy around a massive increase in trade in manufactured goods and foreign direct investment. But this was not a recapitulation of the previous world economy. Abandoning territorial imperialism, 'Western capitalism . . . resolved the old problem of overproduction, thus removing what Lenin believed was the major incentive for imperialism and war' (Calleo 1987, 147). The driving force behind this was the growth of mass consumption in North America, Western Europe and Japan. The products of such industries as real estate, household and electrical goods, automobiles, food processing and mass entertainment were all consumed within (and, progressively, between) the producing countries. The 'Keynesian' welfare state helped sustain demand through the redistribution of incomes and purchasing power. The old 'cross-over' trading system was no longer needed. If before the Second World War the prosperity of industrial countries depended on favourable terms of trade with the underdeveloped world, now demand was stimulated mainly at home.

Moreover, until the 1970s the income terms of trade of most raw materials and foodstuffs tended to decline. This trend had negative effects on the economies of the Third World as a whole, but it stimulated some countries to engage in new models of industrialization which later paid off as they found lucrative export markets for their manufactured goods. The globalization of production through the growth of these Newly Industrializing Countries (NICs), and the increased flow of trade and foreign direct investment between already industrialized countries, finally undermined the geographical production/consumption nexus (often referred to as 'central Fordism') that was the leitmotif of the early post-war decades.

With respect to change over time, American approaches to regulation of the world economy and coordination of economic policy with other countries have gone through a number of mutations. Cafruny (1990) presents one interesting approach to categorizing the principal changes, even though, as he acknowledges, it may be overly schematic (Table 2.7). A phase of 'integral hegemony' is identified as lasting from 1945 until the London gold crisis of 1960 first showed the potential weakness of the gold–dollar exchange standard system (Triffin, 1960). 'Declining hegemony' is dated from 1960 until 1971 when the US government abrogated the Bretton Woods Agreement. The period from 1971 until 1990 is referred to as one of 'minimal hegemony' in which the US has had to face declining rates of economic growth and has had recourse to use of competitive revaluation of the dollar,

Table 2.7 Three phases of US hegemony

	Hegemonic phase		
	Integral (1944–60)	*Declining* (1960–71)	*Minimal* (1971–90)
Hegemonic ideology	conservative/ Keynesian	social democratic/ Keynesian	social democratic/ Keynesian (1971–8) neoliberal (1978–present)
Leadership style	consensual	negotiated	unilateral
Distribution of costs and benefits	United States absorbs short-term costs	mutual	uneven: United States imposes costs
Basic organizing principle	mixed economy/ corporatism	mixed economy/ corporatism	market forces
Basis of élite unity	Atlanticism (collective goods)	Atlanticism (collective goods)	co-option of élites

Source: Cafruny 1990

41

voluntary trade restrictions, and informal methods of economic policy coordination with the main industrialized countries (the G7 summits) to manage its interaction with the world economy.

A vital element in allowing the US to have such a dominant presence within the world economy was the persisting yet historically episodic political–military conflict with the Soviet Union. This had two peaks in intensity in the late 1940s and the early 1980s when each side perceived the other to be increasingly hostile and dangerous (see Figure 2.1). The mid-1970s was generally the period of greatest cooperation.

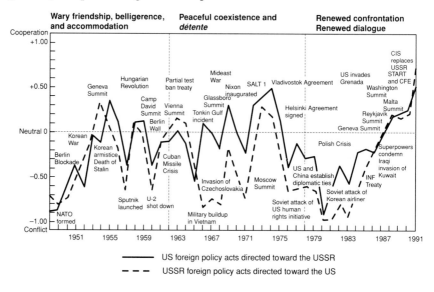

Figure 2.1 The nature of US–Soviet relations, 1948–91
Source: Kegley and Wittkopf 1993, 92

However, even in periods of *détente* or coexistence the overarching Cold War conflict served both to tie Germany and Japan firmly into alliance with the US and to define two geographical spheres of influence at a global scale. For a long time this imposed an overall stability on the world political system, since the US and the Soviet Union were the two principal nuclear powers, even as it promoted numerous 'limited wars' in the Third World of former colonies where each of the 'superpowers' armed surrogates or intervened themselves to prevent the other from achieving a successful 'conversion' (O'Loughlin 1989). For all their weakness, however, Third World countries could not be treated as passive objects of imperialist competition. They had to be wooed and often they resisted. This limited the ability of the super-powers to extend their influence. Unlike in the previous period, the world map was no longer a 'vacuum' waiting to be filled by a small number of Great Powers. The war in Vietnam taught (most of) us that.

The United States was by far the more important global power. One recent study suggests that the *indirect* conflicts of the Cold War (those involving a third party) were overwhelmingly in the vicinity of the Soviet Union (Figure 2.2). Europe was initially the most conflict-ridden world region but later the Middle East and South-East Asia became more important. The United States dictated the terms of the general superpower relationship by defining what was or was not a violation of previous tacit understandings concerning 'interference' in a particular country or world region. Most American setbacks were suffered at the hands of third parties rather than the Soviet Union (e.g. Cuba 1959–62; Vietnam 1965–73; Iran 1979) (Nijman 1992).

A

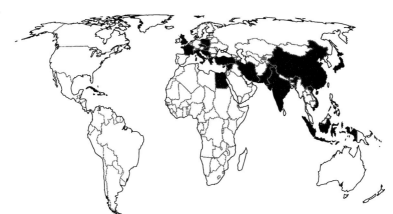

B

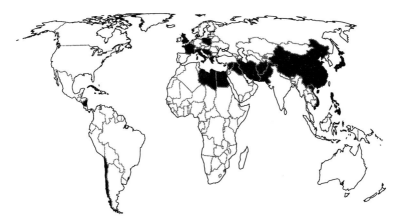

Figure 2.2 The zone of most serious US–Soviet indirect conflict
    A 1948–68
    B 1969–88

Source: Nijman 1992, 688

In the end, too, the Cold War Geopolitical Order came undone with the collapse of the Soviet Union and not America. In some quarters this ending has been greeted as a Hegelian 'end of history' (Fukuyama 1989), with American hegemony in its widest senses – transnationalism, liberalism, democracy – now standing ready to fulfil its (non-)imperial missions abroad. Even parts of the Dark Continent – Africa – are said to be ready to embrace the light of reason and democratization (see Hawthorn 1993a for a discussion). But this conclusion is too self-serving and too simplistic. The Cold War Geopolitical Order came to a formal end in 1989–90 with the break-up of the Soviet Union, but the wider spatial and economic logics that it depended upon had been under attack since the mid-1970s. The Cold War Geopolitical Order was designed at Bretton Woods as well as at Yalta, and by 1980 very little remained of the Bretton Woods system: international investment was growing faster than international trade, the EC and Japan were growing faster than the US, Keynesianism had been replaced by monetarism and supply-side economics, and exchange and interest rates were being set by so-called market forces. The break-up of the Soviet Union was not the only sign of an old order in demise; the Cold War geopolitical *economy* was also in disarray as mounting stagflation, indebtedness and balance of payments disequilibria clearly and successively indicated. We will return to this wider conception of 'disorder' in later chapters (particularly Chapter 7).

CONCLUSION

This chapter has attempted to show how since the early nineteenth century the spatial organization of the international political economy has undergone a series of very specific shifts associated with changes in its structural and institutional attributes. The three geopolitical orders that are identified show distinctive changes in the organization and meaning of space. They are the Concert of Europe – British Geopolitical Order (1815–75); the Geopolitical Order of Inter-Imperial Rivalry (1875–1945); and the Cold War Geopolitical Order (1945–90).

Beyond the particulars of each of these periods, the main concern of this chapter is the conception of geopolitical order. First, in our view spatial ontology must be historicized. The spatial organization of each period differs along two geographical dimensions: the dominant scale of economic accumulation (territorial, interactional) and the dominant space of political regulation (national, imperial, international).

Second, the mutual recognition of states inherent in the state system after 1815 is taken as an important attribute in dating the beginning of the modern international political economy. After this date the world space progressively 'closed' under European domination and a framework arose for inter-state definition of the rules of economic and political competition. The previously ad hoc division of space was thereafter put on a modern 'rational' (rule-governed) basis.

Third, in our usage the 'hegemony' of each period is not bound up by definition with the identity of a particular state. Rather, hegemony refers to the rules and practices which constitute a geopolitical order. These may or may not have a specific territorial agent. Geopolitical order, therefore, does not refer to the apostolic succession of Great Powers to hegemonic (dominant) status but to the changing geographical basis to the international political economy in different historical epochs.

3

GEOPOLITICAL DISCOURSE

Until recently, most geographers and other social scientists would have been satisfied to have identified and offered evidence for geopolitical order and then embarked on an analysis of particular geopolitical conflicts or 'incidents' (e.g. the Anglo-German naval rivalry of the early 1900s, the Cuban missile crisis of 1962, the OPEC oil price increases of 1973 and 1979, the Falklands/Malvinas War of 1982). Geography would have been seen as a set of external facts entering separately into the origins and course of each conflict or incident, e.g. Cuba's closeness to the continental United States, the increased 'dependence' of the industrialized world on 'imported oil', the prospect of oil in the waters adjacent to the Falklands/Malvinas. What would be missing, however, would be an understanding of the nature and origins of the geographical attributions that are implicit in the reasoning of those who brought about the conflicts and incidents. This is now changing as scholars attend to the processes whereby the geopolitical order is represented in the practices of foreign policy.

The purpose of this chapter is to outline a perspective on geopolitical discourse and show how each period of geopolitical order has had associated with it a specific type of geopolitical discourse built in part on understandings first established in Europe during the Renaissance. Even though each discourse is distinctive there are thus textual continuities as old themes are recycled in new contexts. The three discourses or modes of representation we discuss are referred to as civilizational geopolitics, naturalized geopolitics, and ideological geopolitics respectively. Security and economic policies in each of the periods of geopolitical order have been organized around the characterizations of space, places, and peoples defined by these modes of representation.

GEOPOLITICAL DISCOURSE

The term geopolitical discourse refers here to how the geography of the international political economy has been 'written and read' in the practices of foreign and economic policies during the different periods of geopolitical order. By written is meant the way geographical representations are incorporated into the practices of political élites. By read is meant the ways in which these representations are communicated.

Before turning to a review of these representations of space in each period it is important to describe the way in which the term 'discourse' is being used. We do not have in mind the typical idea of 'textuality', whereby a set of texts or documents is scrutinized not for what it may say about practices or behaviour but for its peculiarity, style or 'performative' aspect. Rather what we intend, what could be termed 'discursivity' in general, is related to *context* (Barrett 1991, 126). What is said or written by political élites comes about as a result of the unconscious adoption of rules of living, thinking and speaking that are implicit in the texts, speeches or documents that are produced. But the rules are also constituted in this form as 'epistemological enforcers' (Said 1988, 10) or signifiers to people in general of how they should live, think and speak. The consensus-creating aspect of hegemony in its Gramscian sense, discussed in the previous chapter, is essentially equivalent to this meaning of discourse. But the geopolitical discourses discussed in this chapter are only one part of the complex tissue of discourses (economic, institutional, etc.) making up each of the hegemonies intrinsic to the three geopolitical orders.

There is a danger common to much usage of the term discourse to see it either as a manifestation of thought prior to practice (or activity) or as synonymous with ideology. In the former it is equivalent to an idealism which sees thought as giving rise to practice. Ontology (what there is or what is the case) is confused with epistemology (the limits of human cognitive capacity) to produce a 'world' that is thought rather than both made *and* thought (Norris 1992). In the latter it is seen as a set of ideas either determined by practice or functional to its reproduction. Our usage implies a more tenuous and contingent relationship between thought and practice: that modes of representation are implicit in practice but are subject to revision as practice changes. Spatial practices and representations of space are dialectically interwoven. In other words, the spatial conditions of material life are shaped through their representations as certainly as representations are shaped by the spatial contours of material life.

Another way of putting this would be to say that a discourse is equivalent to a theory about how the world works assumed implicitly in practice by a politician, writer, academic or 'ordinary person'. Even when actors deny they subscribe to a given discourse, careful textual analysis can reveal persisting themes, tropes and a linking genealogy that provides evidence of a discourse that both enables and constrains their practice.

To use the vocabulary adopted in Chapter 1, *geopolitical* discourse involves the deployment of representations of space (Lefebvre 1991) which guide the spatial practices central to a geopolitical order. At times of geopolitical disorder alternative representational spaces vie as candidates to displace currently 'hegemonic' representations of space. In all cases, however, the practical geopolitical reasoning of political élites is the link between the dominant representations of space and the geopolitical order of dominant spatial practices.

The notion of 'political élites' refers to the whole community of government officials, political leaders, foreign-policy experts, and advisors throughout the world who conduct, influence and comment upon the activities of 'statecraft'. From the development of the modern state-system in the eighteenth century there has been a community of leaders and officials engaged in 'foreign policy'. Until the twentieth century this community was small and insulated. But more recently, and especially in the United States, there has been a rapid expansion in the ranks of the 'intellectuals of statecraft', those people specialized in military and foreign policy problem solving. They are now also involved in mobilizing public opinion behind particular strategies and their associated geographical representations.

In summary, discourse is not simply speech, texts or writing but the rules by which these forms are effected. The presence of rules is inferred from the structure, organization and content of texts and speeches. A discourse is not set for all time but adapts to practice. From this point of view geopolitical discourse signifies the rules and conceptual resources that political élites use in particular historical contexts to 'spatialize' the international political economy into places, peoples and disputes.

There are four specific points that follow from these comments on geopolitical discourse and political élites (O'Tuathail and Agnew 1992). First, geopolitical discourse is not simply a separate activity or the identification of specific geographical influences upon a particular foreign-policy situation. Just describing a foreign-policy situation is to engage in geopolitics through the implicit and tacit normalization of a division and description of the world. To identify and name a place is to trigger a series of narratives, subjects and understandings. For example, to designate an area as 'Islamic' or 'Western' is not only to name it, but also to brand it in terms of its politics and the type of foreign policy its 'nature' demands.

Second, geopolitical discourse involves practical reasoning rather than the deliberate deployment of formal geopolitical models; though these often capture important aspects of dominant spatial representations. Practical geopolitical reasoning relies on common-sense narratives and distinctions rather than formal models. Defining areas as modern or backward, Western or non-Western, civilized or barbarian, and democratic or despotic have been important binary oppositions around which modern geopolitical discourse revolves, irrespective of whether its purveyors are academics (disciplinary or formal geopolitics), practitioners of statecraft (practical geopolitics) or media-persons involved in representations in popular culture (popular geopolitics).

Third, the geographical knowledge displayed in geopolitical discourse is usually of a reductive nature. Information about places is filtered and suppressed in order to fit into a priori geopolitical categories. Geopolitical reasoning operates through the active simplification of the complex reality of places in favour of controllable geopolitical abstractions. This is how places and their inhabitants can become 'security commodities', readily subject to invasion, control or bombing.

Fourth, and finally, not all political élites have equal influence over how global political–economic space is represented. Those in authority in the Great Powers or within the hegemonic state (if there is one) have the power to constitute the dominant geopolitical discourse. This happens not only through their own practice but also through the active adoption of the dominant geopolitical discourse by both allies and enemies. Of course, hegemonic representations do not go unchallenged but even challenges often must conform to the 'terms of debate' laid down by the dominant discourse in order to be intelligible or readily understood.

THE ORIGINS AND CONTINUITY OF MODERN GEOPOLITICAL DISCOURSE

Geopolitical discourse at a global scale has its origins in the encounters between Europeans and others during the so-called Age of Discovery. Imperial China and the Arab World certainly had geopolitical schemas for designating the relationship between their political–cultural influence and geographical distance from the political centre. The Roman Empire also made an important distinction between the barbarians outside the *Pax Romana* and the civilized inside. But in none of these cases was the geopolitical distinction seen as fixed for all time or characteristic of a geographical area as such. Conversion and incorporation led to inclusion.

All societies define geographical boundaries between themselves and others (Helms 1988). Sometimes the world beyond the horizon is threatening, sometimes it is enticing. But not all engage in portrayals of the others as 'backward' or permanently disadvantaged if they remain as they are. This is the singular trait of modern geopolitical discourse.

One clear consequence of Columbus's famous voyage of 1492 (and other voyages of the same time) was the heightened sense among European intellectuals over the next two hundred years of a hierarchy of human societies from primitive to modern. The slow 'mental' discovery of America involved a number of attempts at making sense of or ascribing identity to what had been encountered physically (Zerubavel 1992). This process of assimilating the new has led some commentators to see as a founding element of 'modernity' the simple juxtaposition of newly discovered and primitive worlds against a familiar and modern 'old world' from which the discoverers came that simply reproduces itself in the same themes and tropes over the next five hundred years (e.g. Pratt 1992). But why has there been an insistence in modern geopolitical discourse on characterizing geographical difference in terms of a temporal/historical ideal type: modern or backward?

Ryan (1981) suggests one answer when he refers to European intellectuals assimilating the exotic into their own pagan and savage past. 'In the triangular relationship among Europe, its own pagan past, and the exotic, the principal linkage was between Europe and antiquity' (Ryan 1981, 437). The categories of 'pagan' and 'barbarian', discovered as an inheritance from the European Ancients, were deployed to differentiate the new worlds from the Old. Thus, a conception of the temporal transition through which Europe had been transformed was imposed upon the spatial relationship between the new worlds and Europe in its entirety. The religious dimension was especially important in reading the new pagan worlds as standing in a relation to the European Christian world as that world stood in relation to its own pagan past.

This is not too surprising if it is remembered that 'discovery' of the new geographical worlds coincided with the discovery of Europe's own ancient past. Indeed, as Mandrou (1978, 17) points out:

> The new worlds that fascinated the intellectuals of the sixteenth century were not so much the Indies – West or even East – but those ancient worlds which the study and comparison of long-forgotten texts kept revealing as having been richer and more complex than had been supposed.

The texts themselves provided reinforcement to the emerging 'imaginary geography' of modern geopolitical discourse. As Said (1978, 57) suggests, much of ancient Greek drama rested on a rigid separation of Europe and Asia: 'Europe is powerful and articulate; Asia is defeated and distant . . . Rationality is undermined by Eastern excesses, those mysteriously attractive opposites to what seem to be normal values.'

As a result, an image of essential difference with roots sunk deeply in the primordial past was used to invent a geography that had few empirical points of reference (Springborg 1992). In terms of such categories as race, property, oligarchy, aetiology and economy, the Orient (and non-Europe in general) was claimed as 'the negation of all that was being claimed for the West, by polemicists knowing, in fact, very little about it' (Springborg 1992, 20).

As the European states emerged from the dynastic struggles and religious wars of the seventeenth century and embarked upon their schemes of empire-building, comparisons of themselves to the ancient world, and especially the 'model' provided by the Roman Empire, proved irresistible. Lord Lugard (1926, 618), the British High Commissioner of northern Nigeria at the turn of the twentieth century, maintained that Britain stood in a kind of apostolic succession of empire: 'as Roman imperialism . . . led the wild barbarians of these islands [the British Isles] along the path of progress, so in Africa today we are repaying the debt, and bringing to dark places of the earth . . . the torch of culture and progress.' Each of the emerging nation-states of the early nineteenth century could use ancient examples to advantage. Famously, Hegel in the *Philosophy of Right* (1821), on the basis of the extent of the absolute sovereignty of the state and its 'ethical substance', the nation, divided the world into four historical realms arranged hierarchically, with the Oriental as the lowest, the Germanic as the highest, and the ancient Greek and Roman worlds, as the precursors of the Germanic, in between. State-territorial sovereignty and an associated sense of nationhood were the necessary prerequisites for a political entity achieving modern *moral* identity.

Beginning in the late eighteenth century the resort to classical precedent was put on a scientific footing. It became increasingly popular to see social change as a transition from one stage or level of *economic* development to another (Esteva 1992). But it was not until the late nineteenth century that levels of development were seen as the result of environmental conditions or superior environmental 'fitness'. This was often allied with ideas about the intrinsic differences between 'races' and the correlation between the prevalence of different races and levels of political and economic development.

In the late twentieth century the idiom of what Guha (1989, 287) terms 'improvement' has come to prevail over that of 'order'. Geopolitical distinctions no longer lay primarily in essential differences that could be managed but not transcended; they now resided in the possibility that backwardness could be mitigated or even overcome through imitation. Modernity was conceived of as a form of society in which social interaction is rationally organized and self-regulating. In Max Weber's influential account, the rationalization of social life involved the increased regulation of conduct by instrumental rationality rather than 'traditional' norms and values. The version of Weber's theory disseminated by the American sociologist Talcott Parsons makes universal claims. It 'dissociates "modernity" from its modern European origins and stylizes it into a spatially–temporally neutral model for processes of social development in general' (Habermas 1987, 2). But in practice it was the United States that provided the model to which other societies can aspire. The USA defined modernity.

The principal boost to this idiom came from the ideological combat of the Cold War in which the two modern worlds of American capitalism and Soviet communism struggled for dominion in the backward or 'traditional' Third World (Pletsch 1981). The geopolitical polarization of the Cold War was the outcome of a conflict over the meaning of modernity. Its roots therefore lay in competing modern ideological visions. But it was expressed in terms of the essential geographical differences characteristic of all modern geopolitical discourse. The Manichean struggle between the First and Second Worlds was premised upon the existence of a backward world in which each saw its past. *They* are what *we* used to be like. Only by having a backward could there be a modern present and future (Fabian 1983; Duncan 1993).

There is an obvious continuity running through modern geopolitical discourse in the continuing use of a language of difference expressed in terms of a temporal metaphor (modern/backward). However, the idioms and contexts of usage have changed dramatically over time. There is considerable intellectual danger in assimilating all geopolitical discourse into an over-arching continuity flowing from the Renaissance to the late twentieth century and arbitrarily selecting themes or tropes without attention to the contexts (historical and geographical) in which they have arisen. Texts such as those by Pratt (1992) and Spurr (1993), though offering interesting thematic discussions of literary, journalistic and travel genres of writing, tend to choose their examples in a random walk through the history of European colonialism without paying much attention to either its history or its geography. We can and should be more precise.

CIVILIZATIONAL GEOPOLITICS

In the eighteenth century a *civilizational geopolitics* emerged as part of the reaction to the 'struggle for stability' in a Europe that had lost its cosmic centre at the time of the Wars of Religion. This was an element in what Toulmin (1990, 170) has termed the attempt to construct 'a more rational Cosmopolis, to replace the one lost around 1600'. This geopolitical discourse was at its peak of influence in the first half of the nineteenth century. Its main elements were a commitment to European uniqueness as a civilization, a belief that the roots of European distinctiveness were found in its past, a sense that though other cultures might have noble pasts with high achievements they had been eclipsed by Europe, and an increasing identification with a particular nation-state as representing the most perfected version of the European difference.

The idea that the earth's land area is divided into separate 'continents' was first proposed by ancient Greek geographers who identified three continents – Europe, Asia and Africa – bounded by the Mediterranean Sea and the rivers Nile and Don. Although later geographical knowledge suggested that Europe and Asia were not clearly demarcated from one another by a significant body of water, the division of the world into continents persisted because the concept of Europe itself changed. From a physical–geographical region the geographical realm of Europe was transformed into a cultural region. This happened as the Christian Church abandoned its claims to universality and defined a much more narrowly circumscribed Christendom (Hay 1968). The Arab and, later, Ottoman 'perils' gave particular credibility to the sense on the part of vulnerable Europeans of a profound chasm between the familiar world of Christian Europe and the exotic world of the Moslem Other. At the time of the Protestant Reformation in Europe 'The Turkish threat worked toward reviving a waning loyalty to Respublica Christiana and gave new life to the old cry for peace and unity in a Christendom subject to the pope' (Schwoebel 1967, 23).

Imaginative maps showed Europe as a 'queen' among the continents, complete with orb and sceptre (Figure 3.1). This illustrates not only a sense of difference but also an emerging sense of superiority. This was reinforced

Figure 3.1 Europe as a 'Queen'
Source: S. Munster (1588) *Cosmographia*

by the European voyages of 'discovery' which demonstrated the self-evident initiative, vision and zeal of Europeans. Over the next centuries, the feeling of superiority 'gradually hardened into an inflexible conceit that held Europe to be the most civilized and best governed of all the world regions' (Bassin 1991, 3).

The basis for this conceit was not simply comparison with benighted non-Europeans but, more importantly, a sense of a distinctive past. The rest of the world was 'available' for use by Europeans because their history destined them for Greatness. It is no mere coincidence that the early nineteenth century saw an obsession among European élites with the 'examples' of ancient Greece and Rome. This was particularly marked in the United States; in the minds of its leaders very much an extension of European ideals beyond the physical boundaries of Europe. The 'founding fathers' of the US looked to Rome for their model; they thought of Athens as ruled by mobs. In the early nineteenth century this was displaced by a much more popular Greek Revival which affected everything from political oratory to town names and house design. Throughout Europe a 'classical' education became a prerequisite for social and political success. Contemporary events were understood in terms of classical referents.

A commonality giving geopolitical definition to the dichotomy between Europe and the other continents was the frequent combination of a European homeland with a colonial periphery or frontier which characterized such different states as Spain, England, the Netherlands, the United States and Russia. The Other was not so much abstract and unknown as concrete and familiar. At best the political ideas of most European imperialists were that, as in the case of the British in India,

> political power tended constantly to deposit itself in the hands of a natural aristocracy, that power so deposited was morally valid, and that it was not to be tamely surrendered before the claims of abstract democratic ideals, but was to be asserted and exercised with justice and mercy.
>
> (Stokes 1959, 69)

Learning about and understanding the workings of colonial peoples was a vital part of the burden of spreading the light of European civilization. The Nigerian novelist Chinua Achebe (1975, 5) captures the challenge and how it was thought of as follows:

> To the colonialist mind it was always of utmost importance to be able to say: I know my natives, a claim which implied two things at once: (a) that the native was really quite simple and (b) that understanding him and controlling him went hand in hand – understanding being a precondition for control and control constituting adequate proof of understanding.

Unlike the European empires proper, in which the European 'motherland' was separated from its colonies by large tracts of water, in the United States and Russia there was no such clear separation. Peripheries and frontier zones could be identified but there was no obvious physical boundary. The problem was resolved for a time in the Russian case by designating the Ural Mountains as the specific boundary between the European and Asian parts of the empire. Indeed,

> The basic geographical proposition that Russia divided cleanly and naturally into Asian and European sectors entered into the very foundation of the imperial ideology that was refined in the course of the eighteenth century. It was disseminated in the geography texts that began to appear in ever-greater numbers after 1750 and by the end of the century it had become a universally accepted truism.
>
> (Bassin 1991, 7–8)

In the United States, as European settlement spread into the continental interior during the early nineteenth century, the expanding frontier became the root and symbol of the filling out of national territory to the Pacific. Divine Providence was invoked initially to justify the invasion of the continent. The dominant justification, however, became the providential mission of 'America' to spread American ideals and institutions to the Pacific – and beyond. The rhetoric of mission peaked in the 1830s and 1840s when zealots wrote of American 'Manifest Destiny' and 'The Great Nation of Futurity'. In 1847 the US Secretary of Treasury placed in his annual report a section referring to the aid of a 'higher than any earthly power' which had guided American expansion in the past and which 'still guards and directs our destiny, impels us onward, and has selected our great and happy country as a model and ultimate centre of attraction for all nations of the world' (quoted in Pratt 1935, 343).

As this last quotation suggests, the agents of civilizational geopolitics, while invoking a general European outlook and agenda, came from different states which competed with one another as claimants to the mantle of the ancients. 'New' Romes abounded. Yet the claims rested more upon the imitation of French and English models of nation-building than upon some universal shared experience of the 'classical' European past. From successfully policing religious affiliations these prototype 'modern nation-states' had expanded their jurisdictions to construct a set of collective interests, outlooks and traditions.

Particular dynastic and regional identities were reworked as 'national' cultural identities. Classical motifs were combined with local ones. The Napoleonic armies that swept across Europe in the aftermath of the French Revolution left architectural and organizational residues of its classical hubris; from the triumphal arches to the Code Napoleon. But a second influence emanating from England and the German states soon spoke for a less remote past.

Gothic churches, vaulted tombs, museums and houses of assembly, adorned with mementoes of medieval battles and national heroes, filled the lacunae of the nation's collective memory, and little children were taught to revere Arthur and Vercingetorix, Siegfried and Lemminkainen, Alexander Nevsky and Stefan Dusan as much, if not more than, Socrates, Cato and Brutus.

(Smith 1991, 81)

The particular national genius was thereby highlighted against the achievements of a common but ancient past.

Medieval historicism served to foster the sense of membership in singular communities with distinctive attributes that demanded 'self-realization' in 'historic homelands'. The French and English pioneers of nationhood, however imperfect or partial their own achievements, provided important role models for other Europeans, from the Greeks battling the Turks to the Germans and Italians struggling with their historic geographical divisions. Not surprisingly it was to them, and particularly the Britain that emerged victorious from the Napoleonic Wars, that the new nation-states also turned for recognition of their status as modern nation-states.

A clear distinction was drawn between the possibility of this new statehood in Europe and its impossibility elsewhere. Within Europe (and wherever Europeans now lived) nations and states could be conjoined as limited territorial entities that balanced one another mechanically in a system of weights and balances. This analogy to Newton's mechanics provided the image of stability within which foreign policy between nominally equal states was possible. Outside this system was an unlimited space of primitive or decadent political forms that were candidates for conquest rather than recognition. The 'boundary' example of the Ottoman Empire provides a case in point. It was not recognized as a 'member' of the Concert of Europe until 1856 and even then the long-run disputes between it and Russia were never given the attention they deserved. The Otherness of the Turks was a fundamental barrier to their participation in a civilizational geopolitics that drew hard lines around its European homeland.

NATURALIZED GEOPOLITICS

If totemism refers to the practice of interpreting nature in the vocabulary of human groups, such as kinship and genealogy, naturalization does the opposite. It represents the human entirely in terms of natural processes and phenomena. This is what happened to geopolitics from the late nineteenth century until the end of the Second World War. Rather than a feature of civilization, geopolitics was now determined by the natural character of states that could be understood 'scientifically' akin to the new understanding of biological processes that also marked the period. The understanding of states and the geographical organization of the world economy that this gave rise to underpinned the inter-imperial rivalry of the geopolitical order from 1875 to 1945. Naturalized geopolitics had the following principal characteristics: a world divided into imperial and colonized peoples, states with 'biological needs' for territory/resources and outlets for enterprise, a 'closed' world in which one state's political–economic success was at another's expense (relative ascent and decline), and a world of fixed geographical attributes and environmental conditions that had predictable effects on a state's global status.

Naturalization had a number of preconditions. One was the (apparent) separation of the scientific from the subject position of the particular writer or politician. Claims were made to nomothetic knowledge that transcended any particular national, class, gender or ethnic standpoint. So, even as a particular 'national interest' was self-evidently addressed, this was framed by a perspective that put it into the realm of nature rather than that of politics. A vital aspect of this, and another precondition for naturalization, was the conviction that observation of the world's political and economic divisions was a form of innocent perception from which generalizations about resources and power could be deduced. The prefiguring textuality of the geopolitical order and its associated mapping, naming and citing of spaces as 'colonial', 'European', 'powerful' or 'backward' was not often acknowledged.

'Scientifically', in the empiricist sense of the term, the world beyond the immediately familiar was blank and empty but was then filled and labelled according to its varying natural attributes as they appeared from Europe. The world was then known and *possessed* not just politically but epistemologically. This was the great achievement of naturalization; to have depoliticized inter-imperial rivalry into a set of natural and determining geographical 'facts of life'.

One of the most important elements in this naturalized geopolitics was the distinction between imperial and colonized peoples. Citizenship laws, classifications of colonies (particularly between European settler colonies and others), religious missionary activities, and even the emerging division of labour between university subjects (e.g. sociology versus anthropology), operated on this distinction. But this distinction was not merely a 'recognition of necessity', a pragmatic response to an obvious spatial division. Ironically, it rested on the view that (some) Europeans had become masters *of* nature as a result of superior 'fitness' in a natural process of evolution. Transforming or displacing more primitive peoples was justified because as Darwin (1839, 520), one of the intellectual sources for this type of reasoning, had put it: 'the varieties of man seem to act on each other in the same way as different species of animals – the stronger always extirpating the weak.'

In the United States in the 1880s 'scientific' racial claims replaced providential ones as the basis for American destiny. Anglo-Saxons (and other 'Teutons') were identified as exemplars of superior evolutionary fitness. Their superior political principles, growth in numbers and economic power were evidence enough (notwithstanding contradictory evidence in the Census of the United States). Fiske, a leading proponent of the fitness argument, saw an Anglo-Saxon future. 'The day is at hand', he wrote, 'when four-fifths of the human race will trace its pedigree to English forefathers, as four-fifths of the white people of the United States trace their pedigree today [1885]' (quoted in Pratt 1935, 348).

The principle of natural selection thus filtered down from scientific theorizing into popular culture largely in terms of the idea of the 'survival of the fittest'. This became a staple of the journalistic accounts and travel-writing that accompanied the great explosion of European imperialism at the turn of the century. An older idiom justifying colonialism in terms of moral uplift and religious grace gave way to a discourse of racial competition and dominion. Initially invoked to distinguish the leading 'races' of humanity, it quickly lent itself to more refined distinctions, as with the American 'Anglo-Saxons', and served as an important inspiration for the racist ideologies – anti-Semitism, anti-Slavism – and eugenics programmes that flourished politically in Europe and North America in the 1920s and 1930s (Mosse 1980; Yahil 1990; Kuhl 1994). These ideas were shared internationally rather than associated with any one particular country such as Germany.

Jews were especially vulnerable. Not only did the nineteenth century bring rapid advancement to Jews as a group as they moved out of ghettos into society, but some non-Jewish groups 'experienced rootlessness, fragmenta-tion, and estrangement from a once securely anchored, familiar world' (Barnouw 1990, 79). As a result, Jews became objects of hatred and fear to those nostalgic for a world they had lost but without good prospects in the new national industrial society. Demagogues readily portrayed Jews as rootless cosmopolitans in a European world dividing into parochial nation-states. There was no longer a space for difference within the boundaries of states. Jews were dangerous polluters of national homogeneity. Every 'race' was seen as requiring its proper place. In the work of the nineteenth-century geographer Friedrich Ratzel, the founder of the ecological theory of race, the Jews are seen as the one race most 'out of place'.

> In the Near East they were productive (for example, creating mono-theism) but in Europe they have no real cultural meaning. The associ-ation of place and race is linked in the rationale of the German in Africa or the Jew in Europe. They are presented as mirror images, for while the German in Africa 'heals', the Jew in Europe 'infects'.
>
> (Gilman 1992, 183–4)

The Nazi geopoliticians of the 1930s came up with formalized schemes for combining imperial and colonized peoples within what they called 'pan-regions' (Figure 3.2). However fanciful in terms of possibilities of overcoming the political–economic relationships of the time, such mappings did express

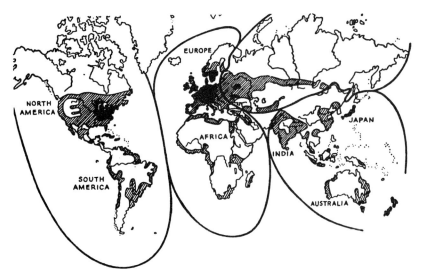

Figure 3.2 The pan-regions of Nazi geopolitics
Source: O'Loughlin and Van der Wusten 1990, 3

in extreme form the common assumption that the world was constituted of racial groupings that could be neatly divided into two 'types' of peoples. The one essentially existed to serve the other. Whatever the precise influence of the Nazi geopoliticians, particularly Haushofer, upon the practice of the Nazi regime (for an argument that it was limited see Bassin 1987), there is little doubt that their ideas fit into a larger political context in which notions of racial hierarchy were conjoined with conceptions of state 'vitality' to justify territorial expansion. The *Zeitgeist* of the epoch even drew into its orbit intellectuals with impeccable anti-naturalist credentials. The classic case in point is the philosopher Heidegger (1959, 39) who wrote in 1935, apparently without tongue in cheek:

> We are caught in a pincers. Situated in the center, our *Volk* incurs the severest pressure. It is the *Volk* with the most neighbors and hence the most endangered. With all this, it is the most metaphysical of nations.

As Wolin (1993) recounts, this geopolitical motif is not merely incidental but very much at the heart of Heidegger's rendering of the 'question of Being'; his central philosophical issue.

At the same time, either through analogy or literally, the European territorial state acquired a status as an organism with its own 'needs' and 'demands'. This too was a transposition from evolutionary biology. But it had older roots. German idealist philosophers such as Fichte and Hegel regarded the state as a being or entity with a life of its own (Benn 1967, VIII, 7). It was but a short step from this to the idea of the state as an organism, a step facilitated by the spread of biological ideas into the emerging social sciences and the rhetoric of politicians in the 1890s.

Like all organisms a state must struggle against the environment (in this case of other legitimate states and 'empty spaces') to survive. This required that it acquire space and resources to feed its healthy growth. The rebirth of militarism in the late nineteenth century (see Chapter 2) was a corollary trend that further fed the image of the state as a permanently embattled entity that could prosper only if individuals and classes subordinated their particular interests to the interest of the larger whole. The doctrine gained expression in terms of three central points: the harmony of state and nation, natural political boundaries and economic nationalism.

The late nineteenth century was a time of tremendous social disruption in Europe. Not only were there massive movements of people within Europe and across the oceans leading to an explosion of urbanization, but increasing capital mobility also undermined the national and local circuits of savings and investment that had previously given an appearance of long-term commitment to the activities of capitalists. These trends were particularly strong in Britain where by the 1890s a reaction had set in against their social consequences. One was the growth of political movements pushing for political and economic rights for workers and women. Another, increasingly popular with political élites, was directed at nipping these new movements in the bud by 'turning the clock back' to a time when all social groups were thought to have lived in local social harmony. Only this time the social harmony was to be realized in the conjunction of the nation with the territorial state.

A mythic medieval community in which each social stratum knew its place and duty to the whole was projected onto the nation as a whole. It was this recreation of the local past in the national present that served to give the state its organic character. An important contrast was drawn between this and the workings of a *laissez-faire* economy. Such an economy was seen as undermining the organic unity of the nation-state. To some influential conservatives it was seen (and still is seen!) as a source of decay, what Spengler was to term after the First World War 'The Decline of the West'. The 'founders' of disciplinary geopolitics, the Swede Kjellen and the Englishman Mackinder, both subscribed to this viewpoint (O'Tuathail 1992b).

This perspective was widely shared by emerging political élites throughout Europe (and its overseas extensions). In Russia the Urals were increasingly discounted as a geographical divide. One influential intellectual current came to see 'Russia as a transcendental geo-historical, geo-political, geo-cultural, geo-ethnographical, and even geo-economic entity, designated by a new generic term: *mestorazvitie*' (Bassin 1991, 16). In Germany local patriotism or identity with the *Heimat* (or homeland) was channelled by conservative political parties into an overarching identity with the *Reich* (Applegate 1990). Only through the advance of the *Reich* could the *Heimat* be defended. In the United States, until the Civil War, 'the United States' was typically a plural noun. Afterwards, it became singular. This transformation is symbolically associated with President Lincoln's famous Gettysburg Address of 1863 when 'What had been a mere theory of lawyers like James Wilson, Joseph Story, and Daniel Webster – that the nation preceded the states, in time and importance – now became a lived reality of the American tradition' (Wills 1992, 145).

Another element in the view of the state as an organic entity was the idea that a state had 'natural boundaries'. This implied, first of all, that the historical boundaries were not necessarily the proper ones. The territorial status quo of the Concert was now called into question. But it also implied that all of the members of a putative nation or ethnic group had a natural right to live within the boundaries of the state. Finally, it also opened up the possibility of using natural features to designate the natural area of the state. Swedish conservatives, such as Kjellen, argued against Norway's independence partly because they claimed that the Scandinavian mountains were not a natural boundary (Holdar 1992, 315). The Nazi concept of *Lebensraum* (borrowed from Ratzel), justifying German territorial expansion in *Mitteleuropa* (and elsewhere) and what was actually called the 'intellectual liquidation' of Poland (Burleigh 1988, 50), had their roots in the notion of natural boundaries (Smith 1980).

This logic was never extended to the 'stateless' colonized world. This is most clearly illustrated in the case of Africa. In the wake of the exploration of Africa, conquest and colonial rule came quickly and devastatingly. In 1884–5 the leading European powers agreed at the Congress of Berlin to stake out their spheres of influence in Africa. The 'Scramble for Africa' which followed over the next twenty years produced lines on a map which had little relation to underlying cultural or economic patterns. Elsewhere the establishment of colonial boundaries was less hasty if often no less arbitrary. These designations continue to haunt these regions to this day (Corbridge 1993).

Lastly, a principal tenet of 'organic conservatism' was economic nationalism. The state was seen as defining the basic unit for economic transactions. Firms and individuals were held to be subordinate to the greater needs of the nation-state. This too was biological in nature. Writers as different in other respects as the English economist Hobson (who influenced Lenin's thinking) and the English geographer–politician Mackinder both shared organic definitions of national interest as the driving force behind economic growth (Kearns 1993). To Hobson empire sacrificed the 'home' economy while for Mackinder empire was the means for maintaining the economic basis to the military power that was essential for national survival. What they and others would have agreed on was the organic unity of the national economy as a 'going concern'. The German *Weltpolitik* (roughly, economic imperialism), though often at loggerheads with doctrines of *Lebensraum* or territorial expansion, nevertheless provided a similar popular logic for acquiring overseas markets and sources of raw materials to underpin German industrial success (Smith 1986).

The idea of a 'closed world' was vital to the plausibility of the language of biology for understanding the European state as an organic entity. As frontiers 'closed' in North America and the world's land masses were incorporated into the world economy, control over territory appeared to be a crucial prerequisite for economic growth. For British élites in particular the certainties of a world economy that worked in their favour seemed to be gone. They were faced both by protectionist rivals and a domestic commitment to free trade that the new geopolitical order had made anachronistic. 'The British economic lead had evaporated, and for the British the world did indeed seem to be shrinking, to be closing in' (Kearns 1993, 29).

But there was more to it than this. From 1880 to 1914 'a series of sweeping changes in technology and culture created distinctive new modes of thinking about and experiencing time and space' (Kern 1983). Such innovations as the telegraph, the telephone, the automobile, the cinema, the radio and the assembly line compressed distance, truncated time and threatened social hierarchies. The global spread of railways and the invention of the airplane were perhaps the most important challenges to conventional thinking about time and space. The sense of a closed world, therefore, was neither illusory nor the product of a uniquely British sensibility.

In a 'closed world' a premium would be placed on relative national efficiency. States must therefore organize themselves to increase their productivity relative to their rivals. No state could afford to rest on its laurels.

> Nationalism and protectionism helped countries mobilize the resources of their earth and their people. . . . A country without access to the full complement of modern industries was vulnerable, would be a pushover in a war, and thus would attract the bellicose attention of more well-balanced nations.
>
> (Kearns 1993, 18–19)

The principal political dispute in late nineteenth-century/early twentieth-century Britain concerned the merits of maintaining the policy of unilateral free trade introduced in 1870. Both sides adopted the language of national interest and 'shared a belief in the importance of "national economic power" but they lacked agreement on exactly what the concept meant or how it should be measured' (Friedberg 1988, 79).

If the liberal economics of the nineteenth century came under attack from the 1890s on, it suffered a nearly fatal final blow from the Great Depression of the 1930s. The mass unemployment of the time produced a variety of intellectual and political reactions (Biersteker 1993). State corporatism pushed the organic analogy to its limits. From this point of view, ascendant in fascist Italy, Nazi Germany, Spain and Portugal, old economic assumptions no longer applied. The national unit was now pre-eminent. State corporatism was 'a system of interest and/or attitude representation, a particular modal or ideal-typical institutional arrangement for linking the associationally organized interests of civil society with the decisional structures of the state' (Schmitter 1974, 86). The state could now be freed from guaranteeing such values as individual liberty and equality to pursue its 'own' agenda of security and foreign affairs with the total support of its economy.

The breakdown of the capitalist world economy also offered an opportunity for some form of socialist internationalism. But Stalin's adoption of 'socialism in one country' in a Soviet 'socialist motherland' effectively reproduced the territorial economism characteristic of other regimes rather than called it into question. This 'model' of socialism – central economic planning and collective agriculture led by a new political élite – did threaten established élites, however, particularly where communist political parties affiliated with the USSR were strong, and led to the association between socialism and 'subversion' that was to be so important in geopolitical discourse after the Second World War.

It was the rationale for government management of the economy provided by the English economist Keynes that provided the most important justification for stimulating the national economy in Britain and the United States in the 1930s and 1940s. Liberalism remained important in both of these countries when it had long disappeared elsewhere. Keynes provided a way of 'squaring the circle', by arguing for a synthesis of private and state activity through 'countercyclical demand management' (Hall 1989). The national economy, however, was the basic unit for Keynesian macroeconomics just as it was for the other political–economic philosophies. Indeed, its basic measures such as national 'propensity to save', national investment and national productivity were indicators of national efficiency that earlier and more illiberal figures would have readily recognized for what they were.

The final feature of naturalized geopolitics was its emphasis on the determining character of geographical location or environmental conditions. The relative success of different states in international competition was put down to absolute advantages of location and to superior environmental conditions. 'Marchland' states were seen by military strategists as possessing intrinsic advantages over 'interior' states because they had fewer contiguous or neighbouring states and, hence, fewer potential adversaries. 'Maritime' or sea-power states were seen as outflanking 'continental' or land-power states in control over the oceans, the main means of global movement. Only the coming of the railway had called this into question; and this because of the relative size or weight of the Eurasian land-mass (or 'heartland') in relation to the difficulty of coalescing and policing the 'insular crescent' (Mackinder) or 'rimland' (Spykman) around its edge. The use of the Mercator projection to represent the earth's land areas served to exaggerate the sense of a world dominated by Eurasia (especially Siberia) because of its systematic enlargement of polarward areas relative to equatorial ones (Figure 3.3).

This spatial–geographical determinism associated with formal geopolitical models, however, was never as popular as a less specific (and more ambiguous) environmental determinism. From this point of view the Great Power potential of states was a function of their industrial prospects which, in turn, could be traced to their natural resources (particularly energy resources) and ability to exploit them. Some went further and claimed that this ability was itself 'determined' by climate. From 1900 until the 1930s such views were not exceptional. Indeed, they formed the mainstream of opinion, particularly among those educated in such academic fields as geography, geology and biology. Much of the academic geography of the period in Germany and the English-speaking world involved elaborating systems of environmental/geographical accounting; classifying states and regions in terms of inventories of resources, racial characteristics, economic and political organization, and climatic types. These were taught in schools and became the 'conventional wisdom' about why some places had 'developed' while others had lagged behind (Fairgreave 1932). The manuals issued to US soldiers during the Second World War are classics of this genre; their title is revealing of their content: *Geographical Foundations of National Power* (US Army 1944). There was no doubt about it, nature determined national destiny.

The Second World War, therefore, was not external to this geopolitical discourse. Naturalized geopolitics pervaded the representations of the war itself in the film and cartographic propaganda in which both sides engaged. On the Axis side it provided the logic for the war and for the anti-Semitic Holocaust that was one of its central features. But the outcome of the war brought to an end the geopolitical order of inter-imperial rivalry and created

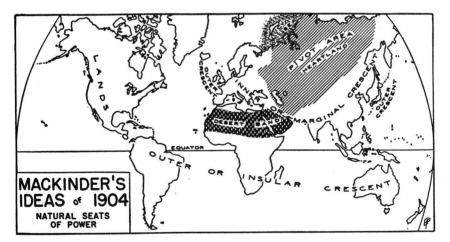

Figure 3.3 Mackinder's 'heartland model'
Source: Mackinder 1904, 424

the conditions for the construction of a new post-war geopolitical order. 'The emerging order was a qualitatively distinctive one, characterized by the breakup of the old colonial empires through the decolonization process, and the emergence of the United States as an economic, military, and political hegemon' (Biersteker 1993, 16).

IDEOLOGICAL GEOPOLITICS

In one sense all geopolitical discourse is ideological, if ideology is defined as an amalgam of ideas, symbols and strategies for promoting or changing a social and cultural order or, as Friedrich (1989, 301) puts it, 'political ideas in action'. After the Second World War, however, geopolitical discourse was centred much more explicitly around competing conceptions of how best to organize the international political economy. Cold War geopolitics was 'linguacultural' (Friedrich's term) more than civilizational or naturalized. By this we mean that the values, myths and catchwords drawn from the experiences of the two victorious states, the United States and the Soviet Union, were to define and determine the terms of geopolitical discourse. One of these, the United States, was to prove the more effective in gaining widespread acceptance for its 'model' of political–economic organization. But its success relied heavily on the active presence of the other as a point of comparison and threat.

Ideological geopolitics has had the following principal characteristics: a central systemic–ideological conflict over political–economic organization; 'three worlds' of development in which American and Soviet spheres of influence vied for expansion into a 'Third World' of former colonies and 'non-aligned' states; a homogenization of global space into 'friendly' and 'threatening' blocs in which universal models of capitalism–liberal democracy and communism reigned free of geographical contingency; and the naturalization of the ideological conflict by such notable concepts as containment, domino effects and hegemonic stability. Older themes from civilizational and naturalized discourses were worked into the new discursive space. The two most important have been the backward–modern polarity and the idea of 'national security'. These will be mentioned as need arises.

The Cold War began as a series of US policies designed to rebuild Western Europe after the Second World War but became a system of power relations and ideological representations in which each 'side' defined itself relative to the other that it was not. This happened in an ad hoc manner but had deep linguacultural roots.

The US and the Soviet Union had been allies during the war. In the aftermath of the war a basic question concerned how Western Europe was to be organized politically and economically. US moves to spread American views of how this might best be done met with resistance from the Soviet Union and its allied Communist parties in Western Europe. From 1945 to 1950 this initial conflict was 'reinforced by a crisis here, a Soviet move there, and an analysis of the protagonists [in the US] which insisted that Moscow was impelled to expand and that only the United States could prevent it from achieving world domination' (Cox 1990, 30). The 'loss' of China to Communist Revolution was a particularly important element in deepening US–Soviet mutual hostility.

But American and Soviet mutual suspicion had origins that went back to the Bolshevik Revolution of 1917. There had been a 'red scare' in the US then that raised the spectre of domestic subversion. US military forces also intervened in the Russian Civil War against the Bolsheviks; the US government knew what it was against. In the 1920s a fear that 'alien' influences threatened the workings of America's unique institutions led to a resurgence of isolationism. 'Foreigners' were dangerous. The United States refused to recognize the new Soviet government in the old Russian Empire (Dallek 1983, 99–109).

Both the United States and the Soviet Union were peculiar states. They both had origins in revolutions with explicit ideological agendas. They both claimed popular mandates that transcended particular ethnic, class or regional interests. They both offered themselves as uplifting lessons in political–economic experimentation to a world where cynicism was rampant. In this context, the practices of 'foreign policy' took on a special importance. As states where what precisely was 'American' or 'Soviet' was unclear or ideational the external threat from things 'un-American' or 'anti-Soviet' became central to American and Soviet national identities. (For a brilliant argument to this effect concerning the United States, see Campbell 1992.)

Discursively, neither was a territorial entity. Each aspired to global ideological dominion. In the Soviet case this was obvious in the ideological lineage that informed official ideology (Marx–Engels–Lenin), if less obvious in the autarkic official practice. In the United States the urge to export the American 'ethos' was longstanding. America itself was an idea that could be sold:

> American traders would bring better products to greater numbers of people; American investors would assist in the development of native potentialities; American reformers – missionaries and philanthropists – would eradicate barbarous cultures and generate international understanding; American mass culture, bringing entertainment and information to the masses, would homogenize tastes and break down class and geographical barriers. A world open to the benevolence of American influence seemed a world on the path of progress. The three pillars – unrestricted trade and investment, free enterprise, and free flow of cultural exchange – became the intellectual rationale for American expansion.
>
> (Rosenberg 1982, 37)

Ironically, given these global pretensions, domestic subversion by foreign agents is a dominant theme in each state's historiography, as is an assumed vulnerability to external threats. Again, the perpetual threat to revolutionary achievements is probably of some significance. In the United States appeals to public opinion have also been of importance in legitimizing any kind of foreign policy. Hence, exaggerating vulnerability and invoking the need for 'self-defence' became an important means of mobilizing public opinion on behalf of foreign ventures (Thompson 1992). The arrival of nuclear weapons (and long-distance delivery systems) led to an American sense of connection to events elsewhere in the world, through the possibility of escalation of local conflicts into a global nuclear one, that previously had been missing. The long history of American territorial security has, however, perhaps left people without much basis for judgement about external 'threats', even as the Soviet experience of invasion and mass murder as recently as the Second World War provided a more obvious basis for collective paranoia.

One important consequence of this shared sense of vulnerability was an idealization by each of the other. Each became a super-potent adversary in the eyes of the other. This is particularly obvious in the American case where there was systematic exaggeration of Soviet economic and military capabilities. For example, Holzman (1989) has shown systematic exaggeration in official US estimates of Soviet military spending from the 1960s to the 1980s. Gervasi (1988) in his annotated version of the Pentagon publication *Soviet Military Power* finds evidence of massive overstatement of Soviet military capabilities that we now know, after the collapse of the Soviet Union, to be absolutely on the mark (Table 3.1). As late as 1988 US foreign policy was still

Table 3.1 The dance of the dinosaurs: US and Soviet procurement of major weapons systems, 1977–86

	United States		USSR	
	Pentagon	Gervasi	Pentagon	Gervasi
ICBMs/SLBMs	850	850	3,000	1,198
IRBMs/MRBMs	200	3,496	1,000	880
SAMs[a]	16,200	84,000	140,000	140,000
Long- and intermediate-range bombers	28	28	375	310
Fighters[b]	3,450	3,450	7,150	2,948
Military helicopters	1,750	2,043	4,650	1,450
Submarines	43	45	90	59
Major surface combatants[c]	89	89	81	44
Tanks	7,100	12,655	24,400	9,370
Artillery	2,750	3,750	28,200	5,625

ICBM, intercontinental ballistic missile; SLBM, submarine-launched ballistic missile; IRBM, intermediate-range ballistic missile; MRBM, medium-range ballistic missile; SAM, surface-to-air missile
[a]Includes naval SAMs
[b]Excludes anti-submarine warfare and combat trainers
[c]Excludes auxilliaries
Source: Gervasi 1988, 121. Gervasi's estimates come from publicly available sources, especially Jane's annual reference volumes on military equipment and Central Intelligence Agency (CIA) annual reports to Congress

based on the continuing existence of a singular Soviet threat to the US position in the international political economy. Even as the Soviet Union disintegrated in 1988–9 a group of leading US experts on military and foreign affairs produced a report that assumed an indefinite continuation of Cold War bipolarity. Though the economic problems of the US – relating to huge trade and federal government deficits – were well known, it ignores them. Instead it engages in statistical and cartographic legerdemain to demonstrate *increased* American vulnerability to the Soviet threat (Figure 3.4 shows one of the maps from the report). It is no exaggeration to say that political élites in each country became so obsessed with each other that, as one of their number put it, he hoped they would not appear to future historians as 'two dinosaurs circling one another in the sands of nuclear confrontation' (Gorbachev 1987/8, 494).

In the context of peculiar 'revolutionary' states with quite different recipes for political–economic organization, abstract terms such as 'communism' and 'capitalism' took on culturally loaded meanings. As the US came to personify capitalism so did the Soviet Union represent communism. Each state became the geographical manifestation of an abstract political economy. Each was foreign and dangerous to the other. The constitutive divide in world affairs after the Second World War thus had specific linguacultural roots that reduced communication to a repetition of buzzwords about the nature of the other party.

Certain domestic interests were served by this geopolitical reductionism. Once under way this reinforced its pervasiveness. In the Soviet Union it disciplined potential dissidents into support of a monolithic state apparatus. In the United States it produced a consensus around a 'politics of growth', an enlarged military economy, and opposition to an 'ideological' politics that would subvert the country from within (Sanders 1983). In other words, it eroded the possibility of an open, competitive democratic politics. The very identity of being American became associated with being 'non-ideological' at home and virulently ideological abroad. 'The simple story of a great struggle between a democratic "West" against a formidable and expansionist "East"' became 'the most influential and durable geopolitical script of [the Cold War] period' (O'Tuathail and Agnew 1992, 190).

As the 'leader' of the West the American President played a key role in giving meaning to the Cold War.

> In ethnographic terms, the U.S. President is the chief *bricoleur* of American political life, a combination of storyteller and tribal shaman. One of the great powers of the Presidency, invested by the sanctity, history and rituals associated with the institution – the fact that the media take their primary discursive clues from the White House – is the power to describe, represent, interpret and appropriate.
>
> (O'Tuathail and Agnew 1992, 195–6)

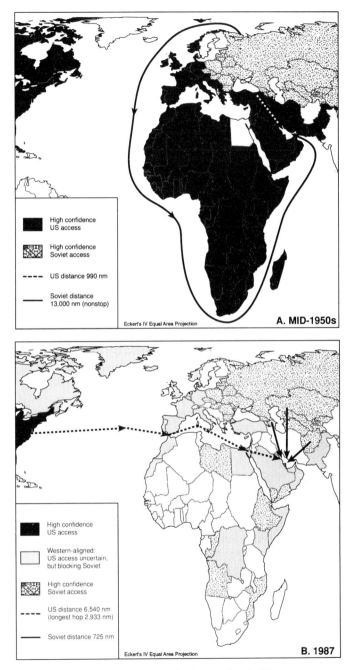

Figure 3.4 How far is it to the Gulf?
Trends in US and Soviet access to airfields or airspace
Source: Iklé and Wohlstetter 1988

69

Through the recycling and repetition of certain images and themes taken from the American past, the President can give a sense of continuity to the practice of foreign policy.

On 12 March 1947 President Truman laid the groundwork for Cold War geopolitical discourse in a speech to the US Congress that while it 'drew the line' against Communism in Greece also integrated this into American history by referring to 'free peoples who are resisting attempted subjugation by armed minorities or by outside pressure'. An American audience knew well that this was a reference to their own experience. Connecting the Cold War with the American Revolution was thereafter a central theme of Cold War discourse. President Reagan, for example, was to regard the Nicaraguan contras ('anti-Communists') of the 1980s as the 'moral equivalents of the [US] founding fathers'.

After the Second World War the US political élite read three lessons from the war: that appeasement of potential aggressors was dangerous, that American national security depended upon the global balance of military forces and not just the balance in the 'Western hemisphere', and that the United States must everywhere oppose attempts at building regional economic blocs or 'pan-regions' such as had emerged in the 1930s. These became the persisting themes of Cold War discourse as they were incorporated into US government practice (Chomsky 1992).

They led inevitably to a 'forward posture' in confronting the Soviet Union. Europe was the main setting for challenging the possibility of Soviet expansion beyond the confines accepted by Roosevelt, Stalin and Churchill at Yalta in 1945. Both the US and the Soviet Union came to share a mutual interest in this scenario. It guaranteed their status as the dominant partners in the alliances they formed in the two halves of Europe. Very quickly, the common goal became the maintenance of balance. 'In this sense the Cold War was more of a carefully controlled game with commonly agreed rules than a contest where there could be clear winners and losers' (Cox 1990, 31).

Although there were episodic uprisings and movements against this spatial division, particularly in Eastern Europe, but also on the part of the Gaullist movement in France, the 'hottest' consequences of the Cold War were felt largely in the 'Third World'. This term is itself a product of the Cold War. Although it was first coined in France in the late 1940s to refer to a possible Third Estate or Third Way, it soon began to refer to those parts of the world outside the settled spheres of influence of the 'superpowers', the First and Second Worlds, in which their global conflict would be concentrated. This logic was to prove visionary. Geopolitical space was conceptualized in terms of a threefold partition of the world that relied upon the old distinction between traditional and modern and a new one between ideological and free. Actual places became meaningful as they were slotted into these geopolitical categories, regardless of their particular qualities.

The threefold categorization turns on a combination of the essential attributes produced by a cross-classification of the pairs of terms 'traditional (backward)–modern' and 'ideological–free' (Pletsch 1981). At a first cut, the modern 'developed' world is distinguished from the traditional 'under-developed' world (the Third World). At a second cut, the modern world is divided into two parts: a non-ideological (capitalist) or natural (free) First World and an ideological (socialist) Second World (see Figure 3.5).

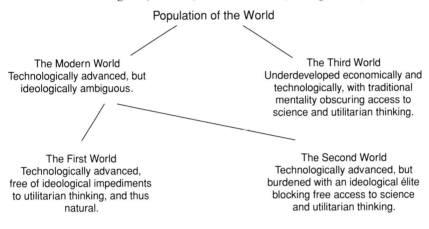

Population of the World

The Modern World
Technologically advanced, but
ideologically ambiguous.

The Third World
Underdeveloped economically and
technologically, with traditional
mentality obscuring access to
science and utilitarian thinking.

The First World
Technologically advanced,
free of ideological impediments
to utilitarian thinking, and thus
natural.

The Second World
Technologically advanced, but
burdened with an ideological élite
blocking free access to science
and utilitarian thinking.

Figure 3.5 The semiotics of the 'three worlds'
Source: Pletsch 1981, 578

The classification has not been contained by its particular origins. Indeed, the concept of the Third World (and Third Worldism) has come to signify resistance to the discursive domination of the superpowers and the possibility of alternative paths to development. For example, theories of development based around the idea of 'dependency' and emanating from such Third World settings as Latin America and Africa, reject the argument that the sources of uneven development are to be found within poorer countries themselves; they lie in their external connections to the world economy (Corbridge 1986; Biersteker 1993, 21–4). For dominant political and academic élites at the height of the Cold War, however, there was little doubt that subordination and imitation were the order of the day. The Third World was that vast geographical zone not yet committed to a particular path to modernity. The success of the superpowers would lie in their relative abilities to recruit candidates for their respective models of political economy from the ranks of the Third World. At the same time, particularly for the Americans, there was the need to counter governments that came to power without an unambiguous commitment to US policies. From Guatemala in 1954 to Granada in 1986 one or other of these rationales was provided for American foreign intervention. (The USSR took a not dissimilar line in Hungary, Czechoslovakia and Afghanistan.)

In each case the local situation was invariably tied to the larger global context. This was because no particular place had singular attributes, only characteristics that followed from its position in the abstract spaces of the Cold War. It was 'friendly' or 'threatening', 'ours' or 'theirs'. This homogenization of global space made knowing the details of local geography unimportant or 'trivial'. All one needed to know was: whose side are they on? One typical headline from a local American newspaper in the early 1970s, the *Columbus* (Ohio) *Dispatch*, said it all. In relation to agriculture in Bulgaria it reported on the wheat harvest of 1973: 'Red Crop Fails'.

Identifying or defining places without a larger classificatory grid is not possible. All maps depend on projections, for example. But 'local knowledge' is possible to a degree that Cold War geopolitics essentially denied. So-called area specialists in government and university departments were frequently frustrated by their lack of influence. The 'big picture experts' like Kissinger, Haig and Brzezinski were always more important than they in policy disputes. Local knowledge was 'background' that could be called on when needed. It did not determine the real identity of places. That was done by putting them into the global frame of reference.

Conflicts with apparent local roots were thus read as local manifestations of the superordinate global one. Links to outside powers in the form of supplies of arms or the furnishing of advisers were read as the only causes of local conflicts. States following their own development or foreign policies were seen as candidates for neutralism or 'Finlandization'; a dreaded condition involving trading with the Enemy but without unequivocal commitment to one side or the other. States were not autonomous actors but agents for one side or the other, expected to fulfil the political and economic goals of the superpowers. Formal organizations such as NATO, the IMF, and the World Bank on the American side and the Warsaw Pact and CMEA on the Soviet side institutionalized the division. In the various United Nations organizations, created at the end of the Second World War to inaugurate a more pacific world order, the two sides expressed their mutual contempt and hostility. The Third World was given a voice but precious little else in UNCTAD (the United Nations Conference on Trade and Development).

The global spatial division recapitulated at a world scale a tendency strong in both American and Soviet (Russian) cultural history to draw strong lines between the space of the 'Self' and the space of the 'Other' (Kristof 1968; Dalby 1988; Campbell 1992; Bassin 1991, 1993). In both cases, if best known in the American case, political discourse has been shaped around the experience of the *internal frontier* which during national development had separated civilization from savagery, domesticated from wild, good from evil.

Another incentive to a neat bifurcation of global space came from the possession of large numbers of nuclear weapons by the superpowers. If allies became involved in conflicts there was always the danger of escalation from local/conventional to global/nuclear war as they drew in their respective 'senior partners'. This served to discipline allies since they provided the places where escalation (i.e. nuclear war) would commence. 'Bipolarity', in the sense of two overwhelming military powers, thus imposed a balance of terror on the world as a whole and froze political boundaries as they were in 1945. But it also restrained the superpowers and, by the 1960s, gave leverage to allies who could exploit the superpowers' fear of mutual destruction once they were able to threaten one another directly with arrays of ICBMs (intercontinental ballistic missiles). Nuclear weapons are not inherently stabilizing even under conditions of bipolarity because there is always a danger of automatic escalation. However, the concentration of nuclear ownership certainly served to highlight the distinctiveness of the two superpowers, whatever their other deficiencies. (See Mearsheimer (1990) for a viewpoint at odds with this and Wagner (1993) for an overview of competing American viewpoints about bipolarity during the Cold War.)

George Kennan, a US official in the Soviet Union at the end of the Second World War, in his famous 'Long Telegram' from Moscow and 'Mr. X' article in *Foreign Affairs* in July 1947, argued that the Soviet Union was a totally alien space with which there could not be meaningful compromise. This claim had an important effect on the policy choices that the Truman administration made relative to the Greek Civil War (1947) and the founding of NATO (1949). Little surprise, then, that the image of huge blocs of space without meaningful internal variation became a fundamental part of Cold War geopolitical discourse (O'Tuathail and Agnew 1992).

Three geopolitical concepts played especially important roles in naturalizing these understandings of space and global politics for Americans and others. These were containment, domino effects and hegemonic stability. Containment, first enunciated by Kennan, referred to the military and economic sequestration of the Soviet Union. Russia's historical geography, not simply its cultural difference, was invoked to give this argument scientific respectability. As Kennan (1947, 574) put it:

> The very teachings of Lenin himself require great caution and flexibility in the pursuit of Communist purposes. Again these precepts are fortified by the lessons of Russian history: of centuries of obscure battles between nomadic forces over the stetches of a vast unfortified plain. Here caution, circumspection, flexibility and deception are the valuable qualities, and their value finds natural appreciation in the Russian or oriental mind.

A policy of 'adroit and vigilant application of counter-force at a series of constantly shifting geographical and political points, corresponding to the shifts and maneuvers of Soviet policy' (Kennan 1947, 575) was necessary to contain the inherently expansive Soviet Union. Throughout his argument

Kennan had recourse to a patriarchal mythology that repeatedly characterizes the Soviet Union as a potential seducer and rapist with repressed instincts that can burst forth at any point along its boundary unless there is constant pressure along all its length to keep it contained (O'Tuathail and Agnew 1992).

Kennan's conception of containment was much more expansive than some later commentators have alleged (e.g. Gaddis 1987; Deibel 1992). But it was at least confined to the margins of the Soviet Union, somewhat after the fashion of Mackinder's 'heartland model', as writers such as Gray (1988) have pointed out. Although there is no evidence of a 'direct' link from Mackinder to Kennan, among some post-war American 'security intel- lectuals' Mackinder's ideas, designed for a very different historical context, took on a prophetic yet scientific role in naturalizing containment as a foreign and military policy (O'Tuathail 1992b).

However, the concept of containment became increasingly expansive after the Korean War (Henrikson 1991; Deibel 1992). The 'geopolitical codes' (Gaddis 1982) of American Presidents underwent a series of shifts from the 1940s to the 1980s. Examining presidential State of the Union addresses for their global geographical content, O'Loughlin and Grant (1990, 527) report the following trend:

> In the 1940s and 1950s, emphasis was placed on the perceived threat to the 'Rimland,' that zone of containment arranged in a semicircle around the Soviet heartland. In the 1960s, attention to specific conflicts in Cuba and Vietnam was added to the dominant U.S./U.S.S.R. competitive theme. In the 1970s decade of detente, attention to foreign policy was reduced, only to be revived strongly in a regional guise in the late 1970s and early 1980s by Presidents Carter and Reagan. During the 1980s, the regional focus of U.S./U.S.S.R. competition has shifted to the Middle East, Southern Africa and Central America.

The main way in which the scope of containment was enlarged was through the logic of the so-called domino theory or domino effects. This argues that the sooner some potential threat to the global status quo was engaged wherever it might occur the less likely was it to produce a spread or contagion effect that could eventually lead all the way back to the United States. In a more sophisticated version the domino theory holds that the credibility of American commitments in more important regions, such as Europe, would be undermined by a failure to protect client regimes in distant corners of the globe. In such circumstances, American resolve to resist any aggression would be opened to doubt and the Great Adversary would be emboldened.

The metaphor of falling dominoes was first used by President Eisenhower in the mid-1950s to describe the consequences of the 'loss' of South Vietnam if the Communist insurgents there prevailed. 'You have a row of dominoes set up, you knock over the first one, and what will happen to the last one is the certainty that it will go over very quickly. So you could have the beginning of a disintegration that would have the most profound influences' (quoted in Gregory 1978, 275). But President Truman had used much the same logic in arguing for American intervention in the Greek Civil War in the late 1940s. His metaphor was different:

> Like apples in a barrel infected by one rotten one, the corruption of Greece would infect Iran and all to the East. It would also carry infection to Africa through Asia Minor and Egypt, and to Italy and France, already threatened by the strongest domestic Communist parties in Western Europe. The Soviet Union was playing one of the greatest gambles in history at minimal cost.

This logic was invoked again at the outset of the Korean War, repeatedly and explicitly during the Vietnam War, and in Chile and Angola in the 1970s. Finally, President Reagan used it as the centrepiece of his administration's policy towards Central America in the 1980s (Slater 1987).

It is a chain-reaction or contagion metaphor, whether in the rotten apple or falling dominoes version. It effectively externalizes local conflicts into aspects of the overarching global conflict. It does so by tying American national security into distant places through the possibility of the spread of revolution or the incubus of Communism. Political change is thereby converted into a natural epidemiological process that threatens to diffuse like a disease into hitherto uninfected regions – eventually back to the United States itself. The fear of appeasement animating the generation that ran American (and Soviet) foreign policy after the Second World War was expressed in a linkage metaphor that put appeasement (or negotiation) beyond the pale of acceptable discourse.

As American dominance over its part of the globe slipped in the 1970s a third naturalizing prop was added to Cold War geopolitical discourse. This one involved arguing that interstate cooperation and an optimal world economy needed a benevolent hegemon, such as the United States, for its best possible operation. 'Both Great Britain in the nineteenth century and the United States after World War II helped bring about an interdependent and overall peaceful world' (Grunberg 1990, 433). Whatever its empirical merit, and this is doubtful, this idea served to 'link the fate of the world with that of the United States, ... there was no escape from chaos, except through a rejuvenation of U.S. power' (Grunberg 1990, 447–8).

Rather like a kindly father or a self-sacrificing hero America's destiny was to hold off the cataclysm that would otherwise follow its demise. The justification for its centrality, however, was different. American images of fatherhood were just too ambiguous! Justification was offered in terms of the *systemic* need for a central provider of 'public goods' – lending at last resort, enforcing liberal trading rules, providing a stable unit of monetary account – that would be underprovided without the hegemon because it would be possible for any one state to consume their benefits without paying for them.

In an anarchic international system only a benevolent despot can provide enforcement in the common interest. Consequently, the successful working of the world economy required continuing American 'leadership'.

Cold War geopolitical discourse, therefore, although overtly ideological also offered various naturalized 'fictions' that could be called up for periodic duty. They allowed the Cold War to become a self-fulfilling prophecy and placed it beyond rational consideration. In general, the articulation of Cold War geopolitics helped to secure and reinforce a set of 'geographical identities ("the West", "the Soviet Union", "the United States"), while serving to discipline domestic social and cultural differences within these spaces' (O'Tuathail 1993a). The polarity between the United States and the Soviet Union was vital to this process. Even though exceptional intellectuals of statecraft, such as President Nixon and his Secretary of State Henry Kissinger, tried to move American foreign policy from its bipolar mode into a multipolar balance of power configuration, their emphasis on demonstrating 'credibility' wherever US power might be in question led them through a domino analogy right back to bipolarity.

The demise of the Soviet Union as an ideological Other has brought the whole discourse into question. Whether ready substitutes will be found is open to dispute. One point is clear, however: ideological geopolitics worked because two states pretending to the mantle of 'modernity' confronted one another globally. Neither a 'militant Islam' nor evil 'drug barons' provide an equally well-defined, competitive, and potent substitute (on Islam see Esposito 1992), although some commentators are now offering demonized portraits of Japan and the Islamic world that would give this impression (see, e.g., Huntington 1993b). Geopolitical discourse must once more be reconstituted (for one attempt using a Mackinderian framework as its guide see Kissinger 1994).

CONCLUSION

Geopolitical discourse, like the political–economic spaces it purports to understand, is not set for all time. In this chapter we have surveyed how (contested) geopolitical discourses have arisen to match the (contested) practices of a geopolitical order. They have provided the rhetorical understandings and dominant meanings through which geopolitical order has been realized in foreign and economic policies. The civilizational geopolitics of the early nineteenth century gave way to the naturalized geopolitics of the late nineteenth and early twentieth centuries. After the Second World War an ideological geopolitics was constituted on the basis of the new geopolitical 'realities' of the time.

Each of these geopolitical discourses has had its own particular combination of elements. But there has also been important continuity. This has been identified in the persisting theme of the geographical projection of 'backward–modern' flowing through geopolitical discourse from its origins in Renaissance Europe. We are living in radically modernist times because we are now critically aware of the Eurocentric and self-serving nature of this way of thinking. A significant challenge is to survey and critically engage alternative scenarios for the future in full knowledge of the workings of past hegemonic geopolitical discourse. Before turning to this task it is important to examine why the field of international political economy has avoided doing so through its elevation of the territorial state into a transhistorical unit of political–economic account.

4

THE TERRITORIAL TRAP

In political theory definitions of the state have two main aspects. One involves the exercise of power through a set of central political institutions. The other entails the clear spatial demarcation of the territory within which the state exercises its power. The former has been uppermost in discussions of state–society relations and the 'relative autonomy' of the state in relation to other putative causes of social life. In international relations theory, however, the second aspect has been crucial. It has been the geographical division of the world into mutually exclusive territorial states that has served to define the field of study. Indeed, the term 'international relations' implies an emphasis on the relations between territorial states in contradistinction to processes going on within state territorial boundaries. State and society are thus related within the boundaries, but anything outside relates only to other states.

The historical development of the relationship between the two aspects of the state has not been well explored. As Walker (1993) has pointed out, political theory has largely concerned itself with 'domestic' politics. Meanwhile in international relations theory the (any) state's essential territoriality has been taken for granted. Much of the literature on international relations (including international political economy) assumes implicitly that a state is a fixed territorial entity (even if its actual boundaries can change) operating much the same over time and irrespective of its place within the global geopolitical order; a state is territorial much like life on earth is terrestrial.

The question of the persistence or the obsolescence of the territorial state has given rise to considerable previous discussion among international relations theorists (the *locus classicus* is Herz 1957). But debate has been overwhelmingly in terms of the presence or absence of the territorial state as an actor rather than in terms of its significance and meaning as an actor in different historical circumstances. This point has been 'lost in endless controversies over whether states are here forever or are about to disappear into some global cosmopolis' (Walker 1993, 14).

Systems of rule or political organization need not be either territorial, where geographical boundaries define the scope of membership in a polity a priori (for example, in kinship or clan systems space is occupied as an extension of group membership rather than residence within a territory defining group membership as in territorial states), or fixed territorially (as with nomads). But the main point of contention inspiring this chapter is that even when rule is territorial and fixed, territory does not necessarily entail the practices of total mutual exclusion which the dominant understanding of the territorial state attributes to it. Indeed, depending on the nature of the geopolitical order of a particular period, territoriality has been 'unbundled' by all kinds of formal agreements and informal practices, such as common markets, military alliances, monetary and trading regimes, etc. (Ruggie 1993, 165), not to mention new systems of telecommunications.

The objective of this chapter is to identify and describe the geographical assumptions that have led international relations theory into the 'territorial trap'. To this end, the first section offers a short discussion of space and spatiality. A second section provides a review of the position taken on the territorial state in the 'mainstream' of international relations. The specific geographical assumptions that underpin the conventional representation are then examined. These are held to define a 'territorial trap' for the field as a whole. A third and final section sketches some recent trends in the world economy that point to the emergence of new spatial forms that the idea of a state territoriality with fixed characteristics cannot adequately capture. These trends are examined in more detail in Part Two of the book.

SPACE AND SPATIALITY IN SOCIAL SCIENCE

The representations of space we use in everyday life to signify our political, social, religious and moral outlooks – left/right, central/peripheral, beyond/within – go largely unremarked. They are unremarkable yet deeply symbolic of how we define what is right and wrong and whom we identify with and against. They are not explicit in the sense of terms we are self-conscious about. A similar situation holds for the conceptions of space and spatiality that have taken hold in various social science fields, including international relations. They are usually implicit or taken for granted rather than openly advertised or contemplated.

In the context of this chapter 'space' is taken to refer to the presumed effect of location and spatial setting – or where political–economic processes are taking place – upon those processes. Spatiality refers to how space is represented as having effects. Apart from a small group of scholars interested in such representations themselves (e.g. Ashley 1988; Weber 1992; O'Tuathail 1993a; Walker 1993), space has been understood most commonly by social scientists in either of two ways. The first sees space as territorial. In other words, space is viewed as a series of blocks defined by state territorial boundaries. Other geographical scales (local, global, etc.) are largely disregarded. This usually taken-for-granted representation of space appears dominant in such fields as political sociology, macroeconomics and inter-national relations. A second understanding views space as structural. From this point of view, geographical entities of one sort or another, nodes, districts, regions, etc. have spatial effects that result from their interaction or relationship with one another. For example, an industrial core area is paired with a resource periphery in a structural relationship of superiority/sub-ordination. This more self-conscious understanding is characteristic of much human geography, economic history, and dependency theories in sociology.

One feature both understandings commonly share is a lack of historical consciousness about the appropriateness of particular spatialities. Rather than a lack of attention to space and a privileging of time, these understandings have idealized fixed representations of territorial or structural space as appropriate irrespective of historical context. In particular, in its attachment to a territorial spatiality international relations theory 'has been one of the most spatially oriented sites of modern social and political thought' (Walker 1993, 13).

The present historical moment has made the nature of spatiality in a wide range of fields an open question in ways that it has not been since the early part of this century (Kern 1983). The dissolution of the Cold War, the increased velocity and volatility of the world economy, the emergence of political movements outside the framework of territorial states (arms control, human rights, ecological, etc.), all call into question the established under-standing of the spatio-temporal framing of 'international relations'.

THE TERRITORIAL STATE
AND INTERNATIONAL RELATIONS THEORY

The merging of the state with a clearly bounded territory is the geographical essence of the field of international relations. The centrality of the association ranges from realist and neo-realist positions where it is vital, to liberalism and idealism where it appears relatively less important. Even that body of work which takes 'geography' seriously, in the sense of adding contiguity or 'regional ecology' to models of inter-state behaviour, usually sees geography as a body of fixed facts setting the environment for the action of territorial states that are essentially the same today as 200 years ago and as much so in Africa as in Europe (e.g. Ward 1990). At least space other than that within the borders of the state gains recognition! But it is important to emphasize that a neo-realist synthesis combining elements of liberalism (the state as equivalent to a rational individual exercising free choice), with a state-centred politics (in which that choice is constrained by the presence of anarchy beyond state borders), has become something of an orthodoxy in American international relations (Ashley 1984; Shimko 1992; Inayatullah and Rupert 1993). One can thus characterize the central tendency in the field as a whole even while acknowledging that differences of emphasis do exist. Even among 'globalist' perspectives (including dependency theories and world-systems theory) only 'critical' international relations theory (e.g. Cox 1987; Gill 1993) regularly avoids the 'territorial trap'.

The importance of the territorial state, and the similar ontological roles (including a fixed identity) it performs within different theories, can be seen in the recent writings of two influential but distinctive theorists: Kenneth Waltz (1979) and Robert Keohane (1984). In examining their writings we do not claim that their work constitutes a 'scientific sample' of all points of view among scholars in international relations. The 'median' would be better represented by Gilpin (1987), particularly with respect to international political economy. We rather claim that they have been especially influential figures towards either end of a continuum of viewpoints in neo-realism stretching from structural realism (Waltz) towards liberalism (Keohane). Focusing on the ends of the continuum establishes a field of disciplinary practice rather than setting up what could be a median 'straw man' alone with his private neo-realism.

Waltz first established his realist view of international relations with the publication of his influential work *Man, the State and War* (1959). In this work he compared three images of the origins of war – human nature, the domestic constitution of states, and the international system. He concluded that the third of these was the best basis for a theory of war. His later work continued this emphasis by focusing on the 'structures of inter-state relations' and totally excluding the internal character of states from the purview of international relations as a field of study. For Waltz the structure of the international system has three features: it is anarchic, without any higher authority; states all perform the same functions and are equivalent units; and there is an uneven distribution of resources and capacities among states. From these fundamental features he draws a number of inferences, in particular that the balance of power is the central mechanism of the international system and that at any specific moment the system's shape as a whole is determined by the nature and number of its Great Powers. Since 1945 the international system has thus involved a bipolar balance of power between two Great Powers in contrast to the multipolarity of the early nineteenth century.

Waltz's (1979) account rests on a 'strong' conception of the territorial state. Unlike classical realist positions, such as that of Morgenthau (1948), Waltz bases his argument on the presumption of international anarchy rather than Fallen Human Nature or the simple desire to master others. In this construction, fear of domination by others, rather than the desire to dominate them, drives inter-state competition (Shimko 1992, 294). This leads Waltz to take the territorial character of the state to an extreme in his claim that international relations should be studied only at the systemic level. This is because it is the anarchy beyond state borders that international relations as a field takes as its subject. The order within state borders is for others to study. From this point of view states are unitary actors whose nature is determined by their interaction with one another. Each state pursues a calculus of status maximization relative to the others. No spatial unit other than the territory of the state is involved in international relations. Processes involving sub-state units (e.g. localities, regions) or larger units (e.g. world regions, the globe) are necessarily excluded. It is a dangerous world out there and if a state (read: our state) is not ready for a competitive environment then it is headed for disaster. This was a reassuringly hard-headed message for Americans during the Cold War!

Waltz's 'system of states' is also structural and ahistorical. If the examples used by Waltz are a guide to his thinking, then antagonistic territorial states have seemingly been around at least since the time of the Greek philosopher–historian Thucydides. For Waltz, the seventeenth-century political philosopher Hobbes stands in direct succession to Thucydides as a philosopher of political realism; a view strongly disputed by many historians of political thought (see, for example, Johnson 1993; Bagby 1994). Waltz pays no attention to the question of the distinctiveness of the modern international system and its roots in the growth of capitalism and centralized military competition in the sixteenth to eighteenth centuries. To retain a parsimonious structural model of international relations Waltz sacrifices historical validity. 'The state system' thus has an existence outside of the historical contexts in which it has evolved.

In contrast, the question Keohane (1984) seeks to address is the orthodox 'idealist' one of how cooperation is possible in the international system without a dominant or hegemonic power. Keohane's book was prompted by growing discussion of the 'relative decline' of the United States and what might replace the 'hegemonic stability' the United States had brought to the world economy. His concern with hegemonic stability is very much of a piece with the ideological geopolitical discourse of the Cold War period. Much of Keohane's argument relies upon the role played by 'regimes' – 'rules, norms, principles, and decision-making procedures' – in governing the institutionalization of the international economy, especially in trade and finance, since the Second World War. In this framework the behaviour of states is not only the result of the competitive pursuit of power in an anarchic world. There are also important incentives for international cooperation; the regimes and formal international institutions that result can significantly restrict state conduct. This is because states are regarded as utility rather than status maximizers. They can gain simultaneously rather than always at one another's expense.

Even here, though, the territorial state remains the central actor. The pattern of international economic relations is seen as determined largely by the policies of states and states' relative economic importance and decline. There is little sense of the state in its 'state-society' aspect or of the political and economic processes within state borders that shape state policies. Moreover, unlike some of Keohane's previous writings in which he argued that non-state actors (such as multinational and global corporations) were eroding the absolute power of states, in this work there is no attention to either the global political–economic system in which individual states are situated or to the shifting balance between state-territorial and other spatial scales of political–economic determination. Perhaps he is accepting the assumption of fixed state territoriality in order to engage with those, including Waltz, who deny the possibility of meaningful cooperation between states?

We would hypothesize that three contextual factors have interacted to reproduce the dominant view about state territoriality found in such apparently different works as those of Waltz and Keohane. One is the preference for abstract and 'closed-system' thinking among advocates of a scientific (positivist) approach to international relations. (On open and closed systems in scientific thought, see Bhaskar 1979.) From this perspective a 'state' is an ideal-type or logical object rather than any particular state and, thus, states can be written about without reference to the concrete conditions in which they exist. If the system of international relations is thought of as an 'open system', such abstract (ahistorical and aspatial) theorizing becomes impossible. Causal chains would form and dissolve historically and geographically. They could not be reduced to a set of primitive terms that would hold true across space and through time. Essential state territoriality is such a primitive term.

A second support has come from the muddling of two terms, state and nation. In the political science literature the term 'nation-state' is often used as synonymous with territorial state. This seems innocent enough, but it endows the territorial state with the legitimacy of representing and expressing the 'character' or 'will' of the nation. Since the nineteenth century the term 'nation' has carried connotations of both cultural singularity and self-governance that boost the transcendental significance of the territorial state when it is associated with them (Doty 1993). Many states are clearly not nations in this singular sense.

A third contextual factor discouraging more dynamic conceptions of statehood and its geographies lies in the intellectual division of labour in political science that emerged in the aftermath of the First World War. The 'international' (meaning inter-state) was theorized as separate and distinct from the national/domestic, and it required a more homogeneous and uniform conception of the state as an actor than that adopted by students of 'domestic' political life (the image of family coziness is shared!). It was also restricted to studying and proffering advice on managing relations between territorial states (see Carr 1939).

Underlying these three contextual factors which have led to the privileging of a territorial conception of the state there have also been three key geographical assumptions. First, state territories have been reified as set or fixed units of sovereign space. This has served to dehistoricize and decontextualize processes of state formation and disintegration. Classical realism and idealism have both relied heavily upon this assumption. But it can be regarded as the 'rock-bottom' geographical assumption that underwrites the others. Indeed, some commentators have restricted their attention entirely to this one (e.g. Weber 1992; Walker 1993). Second, the use of domestic/foreign and national/international polarities has served to obscure the interaction between processes operating at different scales, for example, the link between the contemporary globalization of certain manufacturing industries and the localization of economic development policies. This assumption has been particularly important in neo-realism's fixation on the 'national' economy as the fundamental geographical entity in international political economy. Third, the territorial state has been viewed as existing prior to and as a container of society. As a consequence, society becomes a national phenomenon. This assumption is common to all types of international relations theory.

From an analytic point of view the 'territorial trap' is set from three directions, but operates empirically when at least one of the second two is combined with the first assumption. The first, therefore, is particularly powerful. The assumptions, and how they combine, are now examined in detail.

State sovereignty and territorial space

The concept of 'security' is closely associated in the field of international relations with defending the integrity of the state's territorial space. But this has not signified defence of human, cultural or ecological security, except incidentally. What is at stake is the survival and maintenance of the sovereignty of the state over its territory. The total sovereignty of the state over its territorial space in a world fragmented into territorial states gives the state its most powerful justification. Without this a state would be just another organization. Its claim to sovereignty is what distinguishes the state. Conversely, if there are states then there must be sovereignty (Waltz 1979; Gilpin 1981).

However, the idea of state sovereignty in anything like its modern form is a relatively recent one. It emerged in late medieval Europe in the face of the collapse of the well-established principle of hierarchical subordination (Ruggie 1993). In medieval Europe there were few fixed boundaries between different political authorities. Regional networks of kinship and inter-personal affiliation left little scope for fixed territorial limits. Violence was widespread not because state borders were clearly established but because of frequent switches in political allegiance across fuzzy boundaries (see Fischer 1992; Hall and Kratochwil 1993; and Fischer 1993 for discussion of this claim). Communities were united only by allegiance and personal obligation rather than by abstract conceptions of individual equality or citizenship in a geographically circumscribed territory. Space was organized concentrically around many centres depending upon current political affiliations, rather than a singular centre with established territorial boundaries. In Western Europe the term sovereignty was formally associated with the authority of a monarch (Poulantzas 1980). Time was thought of in terms of repetition of the past rather than an unfolding of novel events in a cumulative or progressive sequence. It was known only as cyclical – seasonal, annual, natural (birth, lifetime, death). The experience of events was not yet associated with a particular 'national' history.

The modern territorial state steadily replaced the plurality of hierarchical bonds with an exclusive identity based upon membership in the common juridical space defined by the writ of the state. In other words, 'the principle of hierarchical subordination gradually gave way to the principle of spatial exclusion' (Walker 1990, 10). The older hierarchical arrangements in Europe involving the Roman Church and the Holy Roman Empire, feudal obligations, and theological claims to just rule, gradually dissolved. Identification of citizenship with residence in a particular territorial space became the central fact of political identity (Sack 1986; Sahlins 1989; Greengrass 1991). Sovereignty shifted from the person of the monarch, identified with a 'divine cosmos', to the territory of the state and state institutions (Collins 1989).

In both political theory and practice the central dilemma became that of the political control of people released from their customary obligations under religious–dynastic authority. The older religious–dynastic communities had relied on a chain of command extending from God through the sovereign and his/her links of political obligation, to the humble subject at the base of the pyramid of power. As this broke down an alternative subjectivity emerged. By and large the new subjectivity involved the emergence of a self-conscious individual constrained in various ways by the rationality of the state and/or the market (a mechanism for coordinating anonymous transactions). Theorists from Hobbes to Rousseau, Kant and Hegel offered different accounts of the new subjectivity. In Hegel, arguably the most important philosopher of the modern state, the individual person was an agent only through the state's definition and enforcement of private property rights:

> capable of owning, acting according to private will, entering into contract, establishing [a] career, accountable legally and morally for stealing from others, and worthy of being held responsible individually for the successes and failures in [their] life.
>
> (Connolly 1988, 118)

In this construction only the state could guarantee the harmonization of society. Only within the homogeneous territorial space of the modern state could the self-conscious subject of modern history emerge. This in turn gave rise to the fundamental distinction between places inside the borders of the territorial state in which 'authentic politics' – the pursuit of justice and virtue – was possible, and the space outside where it was not (Walker 1990; 1993).

Walker (1990, 11–12) gives an eloquent summary of the inside = politics, outside = force logic:

> Inside particular states we have learned to aspire to what we like to think of as universal values and standards – claims about the nature of the good society, freedom, democracy, justice, and all the rest. But these values and standards have in fact been constructed in relation to particular communities. They depend on a tacit recognition that these values and standards have been achieved only because we have been able to isolate particular communities from those outside – an isolation that implies the continuing legitimacy of war and violence.

Politics thus was theoretically possible only on the basis of territorial affiliation rather than non-spatial/categorical identities. This bias was re-inforced by the strong identification with the 'imagined community' of the nation which followed the French Revolution (Anderson 1983). Identities such as those of class or gender, though difficult to organize under modern conditions, have certainly not been absent from modern politics. What has been missing from the accounts of those who link subjectivities directly to the territorial state is any recognition of this. In theoretical terms, such categorical identities undermine the key link between sovereignty and security. Security is only possible for a tightly defined spatial unit endowed with sovereignty. Hence, politics, in the sense of the pursuit of justice and virtue, could exist only within territorial boundaries. Outside is danger, *realpolitik* and the use of force. Security is then, by definition, the defence of a particular spatial sovereignty and the politics within it.

This relationship of security to spatial sovereignty has had four con-sequences for international relations theory. First, it has led to the definition of political identity in exclusively state-territorial terms. This can be seen as progressive in the sense that state sovereignty can involve an active embrace of popular membership or citizenship as opposed to hierarchical subordina-tion by empires, superpowers or multinational corporations (Wolin 1960). However, in the contemporary world there is a remarkable flowering of alternative political identities of a sectoral (gender, ecological, etc.), ethnic and regional character related in part to the threats to 'security' that emanate from changes in military technology, global ecological problems and resistance to repressive bureaucratic government. Increasingly, orthodox thinking about security must engage with shifting sensibilities about political identity (Dalby 1991; Routledge 1993). Conceptions of 'universal' human rights, for example, cynically exploited during the Cold War for political advantage, now are used to call into question conventional definitions of legal sovereignty and the 'right' of governments to mistreat their citizens, shelter war criminals, pollute the global environment or engage in drug trafficking.

In the dominant understanding, geographically variegated, as opposed to territorially homogeneous, forms of political community have been eliminated from consideration by the close association of security with spatial sovereignty. This reflects a persisting tendency in modern European thought to view the autonomy of 'intermediate' or non-governmental organizations as a threat to the vital interests of both individual liberty and the territorial state (Frug 1980, 1089). Indeed, eradicating the power of intermediate groups was for long seen as simultaneously advancing both the interests of individuals and state interests. Powerful subsidiary bodies could be seen as representing a threat to the monopoly of sovereignty exercised by the territorial state. To permit more than one sovereign to function within one territory would create *imperium in imperio*, a dispute over jurisdiction.

It was also assumed that individual liberty could be guaranteed only if there is limited possibility of coercion and interference by other organizations in state 'protection' and regulation of individual citizens. Yet, non-state forms of political community can be defended precisely in terms of their contribution to political freedom and the pursuit of the good life. The arguments of Jefferson and Tocqueville about the stimulus to democracy provided by membership in local primary groups relied upon this point of view.

Second, and also related to a spatially exclusive definition of political identity, is the rigid separation between those people within the territorial space pursuing 'universal' values (politics) and those outside practising different, and nominally inferior, values. This is one dimension of the so-called problem of the Other, in which the people of states other than one's own are represented and incorporated into the world of sovereign states either as 'barbarians' (uncivilized and dangerous but capable of co-existence), 'heretics' (dangerous dissenters who threaten the stability of the system of states and who must be suppressed or converted), or 'primitives' (those who have not yet gained recognition as states and await incorporation into the 'community of nations') (Rosow 1990, 294–9).

In the contemporary world these Others are less easily marginalized than in the past, even as the classic representations of them persist. In particular, massive international migrations, the emergence of middle-level or world-regional 'superpowers' (such as India and China), and the deterritorialization of the communications media combine to limit the confinement of Others in spatial reservations. It has become increasingly obvious to people that the spaces occupied by Others have long been connected to the rest of the world, especially to its Civilized Territories, and have not remained separated or autonomous. However, 'The presumption that spaces are autonomous has enabled the power of topography to successfully conceal the topography of power' (Gupta and Ferguson 1992, 9).

Third, and most misleadingly in terms of the actual 'content' of state territoriality, the security–spatial sovereignty nexus involves viewing the territorial state 'not in its historical particularity, but abstractly, as an idealised decision-making subject' (Ashley 1988, 238). The actual processes out of which different states have arisen are then obscured in favour of an ideal-type territorial state. By way of example, consider the different processes of expansion and incorporation by which the modern states of Britain, the United States and Germany were created. Modern Britain was the outcome of conquest and dynastic inheritance of adjacent territories by a succession of English monarchs over six hundred years working from an initially limited regional base. The United States came about through the settlement of a vast continental tract by immigrants from Europe, Africa and Asia engaged in an expansion that took only about one hundred and fifty years. Modern Germany was created in the mid-nineteenth century by the incorporation of many small German-speaking principalities into a Prussian-dominated *Reich*. Each of these is a territorial state, but each has distinctive origins, geographic scale and founding mythology. Students of comparative politics appreciate this. (For a classic account, see Barrington Moore 1967.) But the peculiar intellectual division of labour in the social sciences has most students of international political economy bundling all of these states together. This has undermined the possibility of seeing the various states as bringing unique contributions in values and behaviour to the system of states, especially in terms of a predisposition towards favouring either economic or military relationships with other states (Rosecrance 1986).

The lack of interest in the creation of specific states has allowed the European confusion of 'state' and 'nation', particularly strong in the English language, to legitimize the idea of a world made up of singular territorial states through a claim to ethnic or national representation by states when in fact most states are not ethnically homogeneous (Ra'anan 1991). The states are the same, it is the nations that differ. It was the success of German and Italian unification in the nineteenth century that confirmed the 'nation-state' as the paradigmatic political unit. This naturalized the territorial state in either one of two ways. The first, notable in the German and Italian cases, was through the dual nature of a nationalism which offered both the universalistic, progressive ideals of the American and French pioneers of the eighteenth century and a collective political identity based on the 'awakening' of an archaic *Volkgeist* (folk-spirit). The second, exemplified by the British, American and French cases, was by the imposition of an ethnic or cultural representation of the national history on an older civic model of statehood.

Fourth, and finally, the principle of state sovereignty 'denies alternative possibilities because it fixes our understanding of future opportunities in relation to a distinction between history and progress within statist communities and mere contingency outside them' (Walker 1990, 14). The only alternatives for political organization as we can imagine it are either continuing division into territorial states or integration into a global superstate. This rests on the view that government through states is necessary because of the axiomatic untrustworthiness of people. 'If men [and women] were to be safe in each other's company, they needed a fundamentally external guarantee of their security – a familiar Hobbesian argument' (Dunn 1979, 23).

But this is not an argument without its own problems. As Locke suggested, the trustworthiness of government remains the larger question: 'This is to think that Men are so foolish that they take care to avoid what Mischiefs may be done them by Pole-cats or Foxes, but are content, nay think it safety, to be devoured by Lions' (Locke 1967 edn quoted in Dunn 1979, 24).

In fact, the territorial state as a primary mode of political organization is no older than the eighteenth century, though it has older roots. This is so despite the best efforts of international relations theorists to find it in the Greece of Thucydides or the Italy of Machiavelli (Garst 1989). The European medieval world was one of local and hierarchical rather than territorial and horizontal allegiances. As late as the Elizabethan period in England, the dominant 'world picture' was still that of 'an ordered universe arranged in a fixed system of hierarchies' (Tillyard 1943, 13). Great power could be exercised by small places; city-states such as Venice, Florence, and Lubeck were world powers. This may never occur again in the way it did before, although the emergence of Hong Kong, Singapore and Kuwait as important financial and industrial centres is suggestive of a move towards 'node and network' forms of political–economic organization. (See Chapter 7 for a sustained examination of these nodes and networks in terms of theories of power and hegemony.) What is more important to note here is that the spatial scope of political organization has not been set for all time and all places in a particular mould. Except during periods in which a balance of power has prevailed in relations between states (as in the mid-nineteenth century) the 'principle' of sovereignty has been routinely violated by Great Powers and other powerful organizations. The sovereign territorial state is not a sacred unit beyond historical time. The state-centring of conventional renditions of international relations assumes precisely that.

The domestic/foreign polarity

Regarding territorial states as the 'nodes' of international political economy, many theorists adopt what can be thought of as a version of abstract individualism (Inayatullah and Rupert 1993). Its theoretical origin lies with Hobbes's world of 'war of all against all' in which territorial states are understood as individuals struggling against one another for wealth and power. In this construction the territorial state is viewed, as in the political economy of mercantilism, as a single, abstract individual: a domestic polity or economy, understood as an identity, in an environment of international anarchy. This is an especially important feature of so-called neo-realist arguments (Milner 1991).

Mercantilism was a loose set of practices and policies followed by many European states in the seventeenth and eighteenth centuries. It was never a coherent economic theory or doctrine. Its most important characteristic was an overt economic nationalism. This was based on the idea that the world's wealth was was basically fixed in size and that, consequently, one state's gain could come only at the the expense of another state's loss. States were thus locked into a permanent and deadly competition for wealth and trading advantage.

In the context of the general economic stagnation of the seventeenth century this perspective had a certain plausibility. However, the view that national economies were the basic building blocks of economic activity in general became strongly grounded in economic and political philosophy. We are its inheritors. The liberal political economy of Adam Smith and his successors, with its emphasis on comparative locational advantage and the logic of rational (self-serving) individual action producing collective optima, has provided a coherent alternative. But it has always been vulnerable, especially during periods of economic stagnation or depression, to charges of depoliticizing resource allocation and devaluing place and social so-lidarities in the service of general consumer or firm welfare (Agnew 1984; Neff 1990).

Under early industrial capitalism the spatial division of labour was strongly organized on a state-territorial basis. The leaders who built many modern territorial states, from Hamilton in the United States to Bismarck in Germany and the Japanese oligarchy of the Meiji Restoration in Japan, all used economic policy to buttress their political ambitions. Over time, however, the increased mobility of capital and the decreased importance of transport costs have produced a global geography of economic activities not readily captured by state-territorial representations of economic characteristics or performance (Massey 1984; Knox and Agnew 1994).

But despite this secular trend in spatial practices, the subordination of the economic to the political in an essentially mercantilist formula is still characteristic of the dominant realist and neo-realist approaches to inter-national political economy. Carr (1939, 116) long ago attempted to ground such interterritorial competition empirically when he asserted, during a very 'abnormal' time in modern history (1939), that 'We have now therefore returned, after the important, but abnormal, laissez-faire interlude of the nineteenth century, to the position where economics can be frankly recog-nized as part of politics.' Carr thus characterized the epoch of inter-imperial rivalry as the normal condition of international relations. This helped to define economic as well as political life as following the fault lines of state boundaries. It also fixed the geographical scale of significant economic and political activities as that of the territorial state. More recently, a recourse to mercantilist ideas has often been to the fore in discussions of the US–Japan trade relationship, in which two territorial economies are often regarded as the leading actors, when in fact its principal features are intraindustry and intrafirm trade and investment rather than territorial competition (e.g. Mastanduno 1991). The historically contingent nature of state–economy relations thus continues to be collapsed 'into a single abstract unity' (Rupert

1990, 429) in which the long-term complementarity of wealth- and power-seeking by territorial states is assumed a priori.

Such intellectual choices eliminate the possibility of seeing the territorial state and its power as dependent on the interaction between global and local (including state-territorial) processes of political–economic structuration. Cox (1981) suggests that territorial states are in a constant condition of reconstruction at the intersection of global and local material conditions. From this viewpoint, showing how the domestic and the foreign come together under different historical circumstances rather than separating them into permanent opposition becomes the overriding task.

We have already addressed this task in Chapter 2, where we outlined three different periods of global geopolitical order for the period 1815–1990. Only in the period of rival imperialisms was mercantilism a dominant feature of international economic relations. Just how misleading the mercantilist position is in contemporary conditions is illustrated by one statistical item that many people (especially economists) spend a good deal of time worrying about, the United States trade deficit (Julius 1990, 81). In terms of the territorial books, the deficit was $144 billion in 1986. If the trading and foreign direct investment activities of US-owned companies abroad and foreign firms in the United States are included in the calculations, however, the huge territorial deficit becomes an ownership surplus of $77 billion. By comparison, Japan's territorial and ownership trade balances are much closer, showing how much more deterritorialized American firms are than Japanese ones (Table 4.1). Most important, an ownership measure of trade raises the basic question of how foreign transactions should be thought of in a non-mercantilist world economy in which perhaps 40 per cent of trade between territorial states is carried on within companies (Reich 1991a).

Table 4.1 Conventional and 'ownership-based' trade balances, US (1986) and Japan (1983), in US$ billions

		USA	Japan
Exports	[1]	224	146
Less intra-firm transfers		123	60
Plus local sales to foreign MNCs		267	3
Plus sales by home-owned MNCs abroad		777	150
Equals total 'foreign sales'	[2]	1,145	239
Imports	[3]	368	114
Less intra-firm transfers		191	65
Plus local purchases from foreign MNCs		445	58
Plus purchases by home-owned MNCs abroad		446	90
Equals total 'foreign purchases'	[4]	1,068	197
Conventional trade balance	[1] − [3]=	−144	+32
'Ownership-based' trade balance	[2] − [4]=	+77	+42

Source: Recomputed from Julius 1990, 81

The competitiveness of many firms in a wide range of industries is now determined by non-territorial factors: access to technology vested in firms, marketing strategy, responsiveness to consumers, flexible management techniques. All of these are the assets of firms not of areas. Large firms grow because of their success in deploying their internal assets. Multinational firms cannot be readily restricted from switching their relatively mobile assets from place to place or state to state. Indeed, territorial states now compete with one another to attract these mobile assets to their territories. In this new world of territory-transcending industry and finance who is regulating whom?

The territorial state as container of society

To the extent that its existence is problematized at all, 'society' in international political economy means what it means in most everyday usage: the social order or organization within the territory of a state. Thus we commonly encounter such phrases as 'Italian society' or 'American society'. This reinforces the totalizing power of the territorial state as a primal force; everything is subordinate to it. Only rarely, as in Bull (1977), is the system of states regarded as analogous to a society. By and large the main contours of society are seen as coincident with the borders of the state.

The historical etymology of the term 'political' gives an important clue to the definitional subordination of society to state. Today the term presupposes the existence of the territorial state. 'That an impersonal structure of domination called the state is the core of politics is an idea so deeply embedded in our ways of thinking that any other conception of it appears counter-intuitive and implausible' (Viroli 1992, 284). Yet this was not always the case. In early Renaissance Italy the term 'political' was intimately associated with society. Politics was 'the art of preserving the respublica, in the sense of a community of individuals living together in justice' (Viroli 1992, 2–3). Only during the sixteenth and seventeenth centuries did politics become the 'art of preserving a state, in the sense of a person's or group's power and control over public institutions (for instance the *stato* of the Medici)' (Viroli 1993, 3). The best known political theorists of Renaissance Italy, Machiavelli and Guicciardini, struggled to reconcile the two but later theorists abandoned the challenge. 'There was not, and there could not be, room for both: either the city of all and for all, or the state (*stato*) of someone' (Viroli 1992, 5).

The close association of society with the territorial state was further reinforced at the beginning of formal sociological thinking in the nineteenth century. A principle of what Smith (1979, 191) calls 'methodological nationalism' came to prevail among a wide range of thinkers. Irrespective of their other profound differences and in degree of its centrality, such figures in the development of modern social science as Durkheim, Weber, and Marx, all shared a territorial definition of 'society'. This is most obvious in such works as Durkheim's *Suicide*, Weber's *Economy and Society* and Marx's (with

Engels) *The German Ideology* (with reference to capitalist if not to communist society). To Durkheim, arguably the most influential of the leading figures called on in American social science, the territorial state was both the creator and guarantor of the individual's natural rights against the claims of local, household, communal and other 'secondary' groups. The state guaranteed social order. But as a 'container' it also provided a territorial unit for the collection of the statistics about social and economic processes that empirical social science required. The categories of the state census came to be the main operational categories of empirical social science.

The intellectual division of labour that emerged in the rapidly expanding universities in the late nineteenth century reinforced this sense of reality. Fields such as sociology, political science and economics were from the beginning concerned with the practical interests of states in, respectively, social control, state management and accumulation of wealth. At their roots they were state-territorial in focus. It is not surprising that derivative fields such as international relations or international political economy should share this orientation.

Reference to local or regional settings, except as 'case studies' of ostensibly state-wide processes, or to 'global' processes, was largely closed off by the 'nationalizing' of social science and its subservience to the territorial state (Agnew 1989). Only outside the modern world, in the 'traditional' societies where states did not yet exist, were other geographical scales of analysis appropriate. This sense of the territorial state as the container of (modern) society has been reproduced in the main currents of international relations. Only inside the state territory is there social order; outside is anarchy and danger.

Mann (1984; 1986) has argued persuasively that a state-centred society has had a definite historical existence but that the pertinence of state-territorial boundaries to what is meant by society is by no means either self-evident or of transcendental relevance. In the ancient Greek polis the nature of the social relationships of the city-state defined the possibility of the pursuit of the virtues (MacIntyre 1981, 152). 'The outer world was only significant if it threatened invasion or promised plunder' (Mazzolani 1970, 16). But this paralleling of social and political organization was unusual. Prior to modern times society was rarely state-defined. But in the twentieth century 'states are central to understanding of what a society is' (Mann 1984, 212). This is because 'where states are strong, societies are relatively territorialized and centralised' (212). 'The territoriality of the state has created social forces with a life of their own' (210). These include:

> the existence of a domestic market segregated to a degree from the international market, the value of the state's currency, the level of its tariffs and import quotas, its support for its indigenous capital and labour; indeed its whole political economy is permeated with the notion that 'civil society' is its [the nation-state's] domain. (210)

The state-defined society, therefore, is 'the product of the usefulness of enhanced territorial-centralisation to social life in general' (211) as groups in civil society (dominant economic classes, churches, military élites) 'entrust power resources to state élites ... because their own socio-spatial basis of organisation is not centralised and territorial' (210–11). However, as Mann is quick to add, 'This has varied considerably through the history of societies, and so consequently have [sic] the power of states' (1984, 211).

This last point is the essential one in the present context. The territorial-state society is a historically and geographically contingent one. In particular, the modern European territorial-state system 'resulted from the way expansive, emergent, capitalist relations were given regulative boundaries by pre-existing states' (Mann 1984, 209). But these relations have never been readily contained. Only when they are contained is the territorial state of central importance in structuring social processes. The territorial state has been 'prior' to and a 'container' of society only under specific conditions. The growth of a strong 'welfare' state and the social changes it allowed, for example, were possible only while capital was relatively immobile beyond state boundaries (Johnston 1993).

The fusion of the territorial state with society, therefore, is not necessarily an intellectual illusion. But what is illusory is its treatment as a Hegelian 'rational unity'. Actual territorial states, based on a circumscribed territory, involve the creation of unified and homogeneous spaces in which various social practices – culture, knowledge, education, employment – are rationalized and homogenized. Making spatial exclusivity is vital to the incorporation of social practices under state regulation. But because space has been subordinated in some instances to the state and became, in Lefebvre's terms (1991, 281), merely 'classificatory' and 'instrumental', the state's spatial unity and internal homogeneity were taken for granted as a 'reality' of social life in general. Lefebvre traces this 'timeless' conception of state-centred space to the influence of Hegelian idealism. Lefebvre (1991, 279) notes that 'For Hegel space brought historical time to an end, and the master of space was the state.' There could be no society without the state. Far from Hegel's immediate influence, indeed in circles that might disavow his every aphorism, this understanding has prevailed. The territorial state unthinkingly serves as the container of society. What better basis for its self-evident importance could there be?

The territorial trap

There is a historiography to these three geographical assumptions. The second two are relatively recent, dating in their current construction from the nineteenth century, even though they have older intellectual sources. They can be thought of as interacting with the older and more fundamental assumption of territorial sovereignty. But they are not simply extensions of it; they have distinctive sources, and they are related to the assumption of state sovereignty in different ways.

In the first place, sovereignty as construed by mainstream approaches implies a relation of similarity among all states in which differences in political and economic practices are defined and demarcated by state-territorial boundaries. The third geographical assumption, therefore, is closely tied to the assumption of territorial sovereignty because the state-society identity is only possible on the assumption of state-territorial sovereignty. This is the way in which the understanding of sovereignty is shared by bureaucratic and popular cultures; practices based on sovereignty such as citizenship, emigration and immigration, policing, trade, national defence and diplomacy are so pervasive that society is easily defined by them (Milliken 1990). It is 'commonsensical' to see the territorial state as the container of society when the state is sovereign.

Territorial sovereignty is also intrinsic to the second geographical assumption, the domestic–foreign opposition. The state's resolution of the problem of order within its boundaries contrasts with the foreign anarchy beyond them. Outside state boundaries there is only struggle for power between the individuals of international relations: sovereign states. There is an essential conflict of interest between states in which one state's gain is always another state's loss unless, in more liberal and idealist accounts, the states can negotiate a temporary regime of cooperation over their antagonistic interests.

The territorial trap, therefore, is circular and cumulative. The geographical assumptions are not linear and additive. They interact to produce mutually reinforcing accounts of international political economy, be they realist, neo-realist or idealist, that are state-centred and in which the space occupied by states is timeless. Theorizing is thus put beyond history by its geographical assumptions.

EMERGING SPATIAL FORMS

The theories of writers such as Waltz and Keohane outlined earlier came to prominence during the Cold War when, one might argue, their orientations to state territoriality had a certain validity. The global conflict between the United States and the Soviet Union effectively froze the world political map into a relatively fixed form for nearly fifty years. During the past twenty years, however, spatial practices, the ways in which space is produced and used, have changed profoundly. In particular, both territorial states and non-state actors now operate in a world in which state boundaries have become culturally and economically permeable to decisions and flows emanating from networks of power not captured by singularly territorial representations of space (Nye 1988; Stopford and Strange 1991).

This dramatic change has led some commentators into speculation about the 'disappearance' of space. Much like at the turn of the twentieth century when Futurists saw the speed of new technologies like the telephone and airplane displacing or homogenizing space, so today there are those who see management of speed replacing control over space. One proponent of this view is Virilio (1986; 1989). Emphasizing the impact of new military technologies on warfare he claims that 'Territory has lost significance in favor of the projectile. In fact the strategic value of the non-place of speed has definitely supplanted that of place . . .' (Virilio 1986, 133). Der Derian (1990) follows this logic in suggesting that with the proliferation of information technologies we can think of 'geopolitics' being replaced by 'chronopolitics' or the spatiality of 'virtual reality' beyond military applications. The whole world is now mastered in a cathode-ray tube rather than on the ground.

We can see what these commentators are implying. The pace of economic transactions certainly has quickened over the past twenty years (Knox and Agnew 1994) and wealth is no longer tied very closely to (national) territory. An interesting example of this is how little of the accumulated wealth of Kuwait was accessible to the Iraqi army after its invasion of Kuwait in 1990. Much of it was stashed away in foreign assets and bank accounts. This illustrates a more general point made eloquently by Luke (1991, 326):

The essentially fictive nature of many contemporary nation-states . . . is exposed by the Kuwaiti and Iraqi experiences in the Gulf War [of 1990–1]. As a classically styled authoritarian state, using modernist myths of military conquest, supreme leadership, national mission, and chiliastic global change, Iraq – like fascist Spain, Portugal, Argentina, Japan, Italy, or Germany before it – demonstrated the bankruptcy of spatial expansion, place domination, and territorial imperialism in the informational flows of contemporary world systems. Kuwait, on the other hand, as a bizarrely postmodern fusion of pre-modern feudalism with informational capitalism, is more of a place-oriented stream within the global flow of money, ideas, goods, symbols, and power. As a point of production and consumption in the flow, however, Kuwait far outclassed Iraq in global significance, even though it had fewer people, less territory, and a smaller military force. . . . Iraq took Kuwait's real estate but failed to capture its hyperreal estate.

This suggests, contra Virilio, that space is not identical to state territoriality. Kuwait has a spatial identity as a node in the network of informational capitalism. This identity now has distinct advantages in terms of garnering and protecting wealth. Kuwait could have others fight for it because of its importance as a node in the emerging global capitalism. But Kuwait still exists grounded in space as well as situated in time. The prophets of homogenization, of time conquering space, confuse state territoriality with space in its entirety.

Be this as it may, the signs of a new spatial organization departing from the conventional spatial representation of state territoriality are everywhere. At one scale there is fragmentation or localization; what Eco (1987) calls 'the return of the Middle Ages'. The Soviet Union, which was in part an attempt to meld many regional ethnic groups into one state, has broken up along its ethnic fault lines. The replacement states are trapped between the desire to acquire the accoutrements of statehood (flags, currencies, militaries, etc.) and the need to collaborate economically with one another. In the former Yugoslavia Serbs and Croats fight violently with one another and with the Moslems of Bosnia over national differences that in a multicultural milieu like New York City would not seem of significant import. Many French-speaking Quebecois openly advocate separation from a state, Canada, that has already given them considerable autonomy. In nominally secular India radical Hindus suggest that the country should become more Hindu, initiating a renewal of regional and religious enmities. In Africa the territorial states inherited from colonialism have failed to establish national identities that override local and ethnic loyalties, leading to suggestions that they are 'failed' or 'quasi' states (Jackson 1990). The position of these states within the global division of labour has worked against the wealth creation that would have led to a displacement of local and ethnic loyalties and to a more effective exercise of sovereignty (Inayatullah 1994).

Regions, religions and ethnicity everywhere challenge territorial states as the loci of political identity. In many countries social classes and established ideologies appealing to 'class interests' have lost their value as sources of identity. Increasingly, the links between the places of everyday life in which political commitments are forged, and the territorial states that have structured them and channelled political activities, are under stress. New loyalties everywhere undermine state political monopolies.

At another scale, in the recent Uruguay Round of the GATT, states negotiated about opening up trade in services, which would involve them admitting more foreigners and 'foreign' ways of doing business into their territories. 'Foreigners' are already migrating at rates rarely experienced in modern world history. In Europe the dominant political issue of the 1990s is the movement towards a more unified European Community (European Union) and whether membership should be expanded or political unification deepened. The world's financial service industries are increasingly globalized, operating around the clock without much attention to state boundaries. Many manufacturing industries have branch plant and research facilities scattered across the globe. Even that most sacrosanct of state powers, the power to wage war, is becoming the mercenary activity that Machiavelli decried in his day. The 1990–1 Gulf War, the first significant post-Cold War war, involved the US in an exercise in coalition building, cost sharing and use of the United Nations that smacked more of collective security with the US as its military arm rather than unilateral action by a single nation-state.

Why have these apparently contradictory spatial forms of fragmentation and globalization emerged together? The most obvious point is that globalization is not synonymous with homogenization (Mlinar 1992; Strassoldo

1992; Lash and Urry 1994). The globalized world economy is based on the transnational movement of the mobile factors of production: capital, labour and technology. As this movement has occurred at an increasing pace localities and regions within states have become increasingly vulnerable to economic restructuring. Previously, during the heyday of the welfare state in Western Europe and North America and state socialism in Eastern Europe, the Soviet Union and China, regional economic policies, national wage agreements and welfare policies, and/or state repression had produced increased equalization across regions within states. With increasing economic competition and increased capital mobility and the collapse of state socialism the outcome has been increased uneven development and spatial differentiation rather than homogenization. Wolin (1989, 16–17) captures the main point most eloquently when he writes:

> Compelled by the fierce demands of international competition to innovate ceaselessly, capitalism resorts to measures that prove socially unsettling and that hasten the very instability that capitalists fear. Plants are closed or relocated; workers find themselves forced to pull up roots and follow the dictates of the labor market; and social spending for programs to lessen the harm wrought by economic 'forces' is reduced so as not to imperil capital accumulation. Thus, the exigencies of competition undercut the settled identities of job, skill, and place and the traditional values of family and neighborhood which are normally the vital elements of the culture that sustains collective identity and, ultimately, *state power itself*. [emphasis added]

One result has been an evolving redefinition of economic interests from national and sectoral (age group, social class, etc.) divisions to regional and local levels. The struggle for jobs and incomes takes place within a global spatial division of labour that is no longer contained by territorial-state boundaries. Another has been that political identities are no longer anchored in singular nation-state identities. For one thing, increasing numbers of people live in what Said (1979, 18) has called 'a generalized condition of homelessness'; a world in which identities are less clearly bonded to specific national territories. Refugees, migrants and travellers are the most obvious of these homeless. But the issue is more general than the question of exile. Globalization is producing a proliferation of multiple (mixed) and local cultures which are not congruent with the boundaries of established territorial states.

From this perspective, globalization has provided the context for fragmentation. Without the first, reducing expectations of and loosening ties to the state, the second, disturbing and reformulating identities, could not occur.

Of course, the territorial state, especially in Western Europe and North America, has continuing strengths within its borders. National political identities are still strong within many territorial states. Mann's (1984) 'society-defining' state is still not exhausted despite attempts in the 1980s to spread the gospels of economic liberalization and privatization of state-provided services. States are principal employers and through their demand for goods and services they are also important economic actors in their own right. The state still provides 'legitimation services' through social spending and most states still exercise a degree of power over economic transactions, not-withstanding an often fragile position within the world economy. States, especially the more powerful ones, are not yet pitiful giants. Labour, investment and, sometimes, monetary policies can still have tremendous impacts on retaining and attracting investment (Parboni 1984; Garrett and Lange 1991).

At the same time, however, states must now mobilize more actively than in the past to attract and keep capital investment within their borders and open up foreign markets to their producers. Much contemporary economic discussion in the United States is about how best to do this. One group preaches a 'geo-economic doctrine' in which the US (and 'its' capitalists) is portrayed as in an economic 'war' with Japan (in particular) (see O'Tuathail 1993b). Another group accepts the advent of transnational capitalism and argues for policies that will encourage investment in the US territorial economy irrespective of its 'national' origin (e.g. Reich 1991a). According to Reich, the question of the day is 'Who is Us?'

Finally, the territorial state has a continuing normative appeal. In his classic work *Politics and Vision*, Wolin (1960, 417) made the case as follows:

> To reject the state [means] denying the central referent of the political, abandoning a whole range of notions and the practices to which they point – citizenship, obligation, general authority, ... Moreover, to exchange society or groups for the state might turn out to be a doubtful bargain if society should, like the state, prove unable to resist the tide of bureaucratization.

However, such a juridical state should not be confused with the absolute sovereign state of conventional modern political theory. Territorial states as we have known them are not necessarily the best instruments for Wolin's political life. In support of a 'cosmopolitan ideal' of democracy, for example, Pogge (1993) advocates a world-wide multilevel scheme of political units to encourage a 'vertical dispersal of sovereignty'.

The main point in reviewing the continuing strengths of territorial states is to suggest that globalization and fragmentation do not signal their terminal decline; the Final Fall of the territorial state. But at the same time, and this is the main point of the chapter, the world that is in the process of emergence cannot be adequately understood in terms of the fixed territorial spaces of mainstream international relations theory (and international political economy).

CONCLUSION

By means of three geographical assumptions the terrritorial state has come to provide the intellectual foundation for the mainstream positions in international relations theory – realist, neo-realist and liberal. The first assumption, and the one that is most fundamental theoretically, is the reification of state-territorial spaces as fixed units of secure sovereign space. The second is the division of the domestic from the foreign. The third geographical assumption is of the territorial state as existing prior to and as a container of society. Each of these assumptions is problematic, and increasingly so. Social, economic and political life cannot be ontologically contained within the territorial boundaries of states through the methodological assumption of 'timeless space'. Complex population movements, the growing mobility of capital, increased ecological interdependence, the expanding information economy and the 'chronopolitics' of new military technologies challenge the geographical basis of conventional international relations theory.

The critical theoretical issue, therefore, concerns the historical relationship between territorial states and the broader social and economic structures and changing geopolitical orders (or forms of spatial practice) in which these states must operate. It has been the lack of attention in the mainstream literature to this connection that has led it into the territorial trap. In the second part of *Mastering Space* we will comment further on the dangers of hypostatizing the territorial state at a time of rapid deterritorialization of the global political economy. In so doing we will also comment critically on two more sets of concerns that remain central to international relations theory: the view that because some territorial states have succeeded one another in the past as Great Powers, so another Great Power must succeed the US as the next hegemon; and the view that the world system must tend to instability and disorder in the absence of a singular hegemonic Great Power.

Part II

Hegemony/Territory/ Globalization: the Geopolitics of International Political Economy

5

'HEGEMONIC' INSTABILITY AND THE RELATIVE DECLINE OF THE UNITED STATES

Since the 1970s the Cold War Geopolitical Order has slowly unravelled. It was the erosion of the American rather than the Soviet position which first attracted attention. Many students of international political economy trace the initial breakdown to the collapse of the post-war international monetary system (Bretton Woods) in 1971–3 and the dramatic increases in world oil prices in the decade of the 1970s. An overwhelming emphasis has been placed on the idea of American economic decline compared to the pre-eminent position of the US during the 1950s and 1960s. This, along with the more recent collapse of the Soviet Union, is seen as having opened up the possibility for a new geopolitical order in which the United States will no longer be the hegemonic power or 'hegemon'. A corollary of this trend is thought to be the rise of a new hegemon. This makes the question of American decline both more threatening and more urgent. American hegemony will be displaced by a hegemony exercised by another territorial state.

The logic of previous chapters was that there can be hegemony in the international political economy without an identifiable territorial hegemon. In this chapter, however, we take seriously claims concerning hegemonic succession and stability in relation to the contemporary US. A first section outlines the concepts of hegemonic succession and hegemonic stability which have provided much of the basis for the debate about relative decline.

HEGEMONIC SUCCESSION AND HEGEMONIC STABILITY

Underlying arguments about American relative decline is a deeper concern for questions of hegemonic succession and stability. The key claim of hegemonic succession theories is that every 'hegemon' (or dominant power) must finally decline as more efficient competitor states overtake them. This was the fate of the Netherlands in the eighteenth century and Britain in the early twentieth century, and now it is to be the fate of the USA, which will cede some of its powers to a more efficient successor (or successors). This is the logic which drives Paul Kennedy's (1987) account of the rise and fall of the great powers, and it is apparent in different guises in the work of Gilpin (1987), Modelski (1987) and Wallerstein (1984). From this point of view, modern world history is written around a succession of hierarchical systems or 'hegemonies' in which a particular territorial state is identified as the rule-giver and enforcer for the period in question. In contrast, theories of hegemonic stability focus only on the periods of British and American hegemony. The commitment of these two countries to free trade is singled out as an instance of the benefits conferred on the international community by a hegemon willing to enforce a set of transcendental rules even when this hurts its own prospects. Some proponents of this position go even further, arguing that the United States differs profoundly from all other states. Kindleberger (1976; 1986) sees the United States as a benign leader concerned for its 'conscience, duty, [and] obligation' (Kindleberger 1986, 845) to the world at large. American altruism provides the key to understanding how the Cold War Geopolitical Order worked and why the United States is still needed as a leader if the world is to experience renewed cooperation and harmony.

Neither of these arguments is very secure in the forms they are conventionally presented. We have already argued that hegemonic succession theories depend on a view of history as repetition, in which a problematic past is allowed to serve as a guide to an unproblematic future. These arguments fall prey to the territorial trap we described in Chapter 4. For their part, hegemonic stability theories do contain some useful insights, but the empirical propositions that they rest upon have been challenged in three important respects. First, since 1789 only about sixty out of over 200 years can be associated with a 'definitive' hegemon: 1845–75 with Britain, and 1945–73 with the United States. Other sorts of arrangements have been more typical: balance of power, territorial empires, inter-imperial rivalry, or combinations of balancing and hegemony. Second, during their periods as hegemons Britain and the United States were not unmitigated sponsors of free trade. Free trade was rather part of their strategy to gain advantages in trading high-value goods with certain partners. Elsewhere they preferred formal empire (Britain) or managed trade (the US). Third, the benefits of hegemony accrued overwhelmingly to the hegemon. Calleo (1987) has shown how both the *Pax Britannica* and the *Pax Americana* were characterized by self-serving behaviour on the part of the respective hegemons at the expense of the system as a whole. For example, the gold standard system allowed Britain to drain funds from the periphery of the world economy as well as regulate the world monetary system (Eichengreen 1988); the Bretton Woods system of exchange rates (1947–71) allowed the United States to finance its balance of payments deficits at the expense of other countries (and currencies).

Why, then, have hegemonic succession and stability theories continued to dominate the literatures of international relations and international political economy? We can think of two main reasons. On the one hand, it is clear that these theories satisfy a number of intellectual needs felt in the United

States (and elsewhere) since the 1970s. One need expressed by both theories is that of linking together political forces with economic outcomes at a time when there has been a spreading perception of American decline relative to certain rivals. Two other needs find expression in hegemonic stability theories. These are the need to tie together the fates of the US and world economies in a seamless web of mutual necessity, and the need to identify the United States as a benevolent leader that is still 'bound to lead'. The first of these involves a systematic exaggeration of the dependence of the US economy on international trade. The US has a huge 'internal' market, certainly by comparison with such putative previous hegemons as the Netherlands and Britain, so an emphasis on national economic 'competitiveness' does not make much sense in purely economic terms (Krugman 1994). What it does do is provide an alternative external stimulus to the demise of the Cold War. The second need suggests that the US can still be the 'winner' in the global 'economic market-place' that has replaced the Cold War as the 'playing field' for inter-state competition.

On the other hand, and in still wider terms, this list of needs satisfied by hegemonic succession and stability theories reminds us of the Pandora's box analogy made by Grunberg (1990). Grunberg traces the popularity of hegemonic stability ideas to a 'powerful mythology' in American (and Western) culture that personifies states as either predatory or self-sacrificing. Once in place such personifications become second nature and difficult to 'refute' empirically. The implication is that it is better the devil you know (American hegemony) than the devil you don't. As Grunberg puts it: 'This is the "Pandora's box" effect; Pandora was, of course, a pagan version of Eve' (Grunberg 1990, 473). Who wants to mess with her?

More prosaically, hegemonic stability and succession ideas continue to receive widespread support because they appear to be speaking to the real concerns of economic and political actors in the United States and elsewhere. As compared to the 1950s and 1960s, the US territorial economy is not in such good shape, and there are now seeming rivals to US economic and financial powers in the form of Germany, a wider EC, and Japan. An ostensibly empirical debate on the relative decline of the US economy intersects forcefully with a theoretical literature on the rise and fall of Great Powers.

This chapter intervenes in the debates about hegemonic succession and stability by considering in detail the question of US relative decline. (The next chapter considers whether there is a plausible pretender to the US throne.) The next part of the chapter reviews seven linked arguments on the so-called relative decline of the USA. We follow this with a critical account of those arguments which call for – or foresee – a continuing American leadership role. A final section offers a perspective on hegemonic instability that separates the issue of American hegemony from the condition of the US territorial economy. This returns us to two of the main themes of Part I of this book: the nature of hegemony and the dangers of the territorial trap.

THE RELATIVE DECLINE OF THE UNITED STATES

The topic of America's relative decline as a hegemon has attracted considerable attention from a large number of scholars. It is impossible to do more here than briefly review the evidence for and against seven main perspectives on the topic. All of these perspectives examine American decline relative to past US performance and the performance of other states (especially Germany and Japan). The perspectives differ, however, in the core problems they identify (see Table 5.1). Two perspectives are macrohistorical in form, seeing the American experience as the latest manifestation of a persisting historical pattern, while five perspectives remain more or less entirely concerned with the American case. (Of course, these are generalizations; all seven perspectives overlap in part and make reference to a variety of economic, social and political factors in their explanations.) Reading Table 5.1 top-to-bottom, it will also be apparent that four perspectives identify a profits squeeze (in various guises) as the core problem facing American businesses, while three perspectives identify national (especially public) spending as the villain of the piece.

Leading sector decline

Advocates of a 'leading sector decline' perspective reserve a prominent position in their explanations for technological innovation. Hegemons arise because they are more successful than other states in capturing the expanding profits of new technologies in 'leading sectors'. However, they remain overcommitted to these and as new leading sectors arise other states can cash

Table 5.1 Relative decline arguments

	Profit squeeze	*National spending*
Macrohistorical	Leading sector decline (e.g. Wallerstein 1984; Thompson 1990)	Imperial overstretch (e.g. Kennedy 1987; Gilpin 1987)
American Case	1 Capital–labour shift (e.g. Bowles *et al* 1983; Castells 1980)	1 Crowding out (e.g. Friedman 1988; Thurow 1985; Calleo 1992)
	2 Accumulation crisis (e.g. O'Connor 1981; Cox 1987)	2 Investment to consumption shift (e.g. Sommers 1975; Brittan 1975)
	3 US–world economies identity shift (e.g. Wolfe 1981; Agnew 1987b; Reich 1991a)	

Source: After Rasler 1990

106

in at the expense of the hegemon. Advocates of this position differ in the relative centrality they give to rates of profit, reinvestment, prices of primary commodities and war in bringing about the redistribution of the fruits of technological innovation (Figure 5.1). While there is no consensus on the direction of causality, however, there is agreement on the search for historic-ally invariant patterns of causation (Thompson 1990, 204–5).

What should we make of this perspective? There is empirical evidence for a certain similarity in the slowdown of American leading sector growth in the 1980s with the British record in the 1890s, but the precise determinants in each case have not yet been delineated. What is clear is that the US lead in new technologies has declined precipitously since 1960 (O'Loughlin 1993). The ratio of new patents awarded to foreigners increased from less than 10 per cent in 1966 to nearly 50 per cent in 1989. The biggest jump was around 1980 and by far the largest proportions of the new patents have been awarded to Japanese and German firms and individuals.

Set against this, however, we should note that certain changes in circum-stance have conspired to make analogies with the past dangerously misleading. First, the pace of the diffusion of technological innovations today is much more rapid than in the past. It is now more difficult for particular states to capture the singular benefits of new technologies which are often the property of firms that operate on an international basis. 'Leads' become shorter and shorter as technologies diffuse more quickly. As Thompson (1990, 232) concludes in his useful survey of 'leading sector' arguments: 'there is no reason

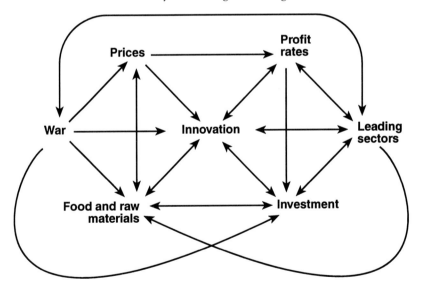

Figure 5.1 Potential causal relationships involved in long-waves of leading sector development
Source: Thompson 1990, 204

to assume that the leading sector wave must continue.' The 1990s will be a decisive decade since this is the period in which leading sector arguments agree there should be a dramatic technological upswing dependent on biotechnology, computers, robotics, lasers and information technologies.

These new technologies are particularly disruptive of the territorial partitioning of space on which state capture of the benefits of technological change rests. They encourage the direct interaction of localities in global networks and flows, without the necessary mediation of national modes of production. As Reich (1991a, 172) puts it: 'Barriers to cross-border flows of knowledge, money, and tangible products are crumbling; groups of people in every nation are joining global webs.'

Second, as much of the most recent research on long-waves in the world economy suggests, the dynamics of technological change can be expected to evolve as the institutional conditions that characterize different periods change (Perez 1983; Tylecote 1992). From this viewpoint, there need be no expectation of spatially clustered technological change driving hegemonic decline and succession. That this did happen in the past is no guarantee that it will re-occur. In the American case the dominance of big firms in a relatively uncompetitive domestic environment, and the favouring of military-orientated research by the federal government, probably have had more to do with the changing national patent balance mentioned earlier than does some kind of national intellectual and organizational exhaustion. With the growth of an increasingly competitive American market due to foreign pressure and the end of the Cold War the conditions for technological development in the United States have changed considerably. The US is now well-placed to exploit the new informational technologies. It has the world's best telecommunications infrastructure and a higher rate of absorption of personal computers than other leading countries, including Japan and Western Europe (Table 5.2).

Table 5.2 The US, Japan and the European Community in information technologies

	USA	Japan	Europe
Telephones			
Lines per 100 people	48.9	42.2	42.2
Calls per person per month	43.4	46.1	48.7
Cellular telephones per 100 people	2.6	1.2	1.2
Television			
Households with cable (per cent)	55.4	13.3	14.5
VCR-related expenditures per household per year in dollars	44.6	35.3	14.1
Computers			
Personal computers per 100 people	28.1	7.8	9.6
Data-base production (per cent of world)	56.0	2.0	32.0

Source: Consumer Federation of America

Imperial overstretch

A variety of intersecting perspectives emphasize the overriding importance of military spending and foreign military commitments on the long-term investment and productivity characteristics of hegemonic states or 'system leaders'. The basic logic shared by these perspectives is that, while the expense of military preparations is initially within the economic capabilities of the leading state (or states), military commitments tend to increase over time and eventually undermine that country's rate of economic growth. Other states without extensive military commitments can invest more in their economies and thus overtake the hegemon in terms of economic capabilities.

The imperial overstretch thesis in its fully-fledged form has been popularized by Kennedy (1987), but particular aspects of it, such as the drain on the balance of payments from foreign commitments and the diversion of expertise and capital from the civilian to the military economy, also figure as 'exacerbating' factors in many other explanations of relative decline. Calleo (1992), for example, incorporates the balance of payments drain, and military spending in general, into his argument that national government spending in the United States has 'crowded out' more productive investment in the economy.

The evidence for a structural defence spending–investment substitution effect across a number of Great Powers (including the United States) in different historical periods is relatively weak. Rasler and Thompson (1988) found that only France (1872–1913) and the United States (1948–78) demonstrate the trade-off effect assumed in the imperial overstretch model. There is no evidence for a generic effect across a set of states including Britain (1813–1913; 1946–78), the US (1871–1913), France (1821–1913; 1951–78), Germany (1873–1913; 1951–78), and Japan (1886–1913; 1946–78). The absence of evidence for a generic effect, which would have been supported more strongly if the nineteenth-century British case had emerged as significant, suggests that the defence–investment trade-off is not inevitable for either 'system leaders' or other Great Powers (see also Offer 1993).

Even in the American case (1948–78), moreover, the addition of variables which measure private and public consumption, deficits, savings, productivity and economic growth, renders the simple trade-off argument open to question. According to Rasler

> previous fluctuations in consumption (both private and public, military and nonmilitary) and deficits do not correlate significantly with previous fluctuations in investment, productivity, and economic growth from 1950 to 1986. Such evidence points to the verdict that there are larger, underlying forces of stagnation that predate the observable downturn in U.S. productivity and economic growth in general.
>
> (Rasler 1990, 186–7)

These results are hardly definitive, but they do cast a shadow over those narrative accounts which never rigorously confront their empirical claims.

At the very least, more appears to be bringing about American relative decline than excessive military spending as part of a historic pattern of hegemonic decline.

The five remaining explanations are all restricted to the American case. They all concentrate on the investment–productivity linkage and in theory could be applied to any national economy. They presume a particularly strong connection between capital investment and productivity, although workforce characteristics, technological innovation and industrial relations also clearly affect productivity. The US productivity 'slowdown' is traced either directly to the working of the economy or indirectly to the policies of the national government. It is important to note that each type of explanation also finds some role for the factors identified as central by the others.

Profit squeeze 1 – capital–labour relations

The first of the profit squeeze theses identifies the changing nature of capital–labour relations within the US economy as fundamental to explaining declining investment and productivity. From this point of view, after the Second World War the US labour force benefited from the growth of big business in the form of higher incomes. These incomes in turn stimulated consumption and expanded profits. However, large monopolistic firms failed to engage in sufficient investment in new product and new plant development. Built-in wage increases for many workers also cut into profits. As a greater proportion of the national income was redistributed from capital to labour there was less available for investment and future economic growth.

The main problem with this perspective is that since 1974 the real incomes of the American labour force have stagnated (Cutler and Katz 1992; Levy and Murnane 1992). At the same time the rate of profit in the American economy has turned upwards. But this has benefited neither investment nor productivity. There has been a redistribution of income from labour to capital without any discernible positive effects on the economy as a whole (Friedman 1988). Investment abroad and increased consumption of imported goods appear to have been the two principal consequences.

Profit squeeze 2 – accumulation crisis

The accumulation crisis thesis rests on the view that a national economy based on private ownership requires considerable state mobilization of resources in maintaining economic growth (investment in infrastructure, education, healthcare, etc.) and in supporting social harmony (unemployment benefits, welfare payments, etc.), but that paying for these demands increasingly cuts into the profits of businesses. A fiscal crisis results when the state maintains its spending at levels beyond what its revenues can support. Political interest groups with influence over politicians refuse to see their programmes cut and the government is unable to finance new policies to improve productivity. The end result is an economic crisis.

The problem here is that the picture composed by this perspective is one common to many contemporary states. Italy and Sweden, to name just two states, face fiscal crises in the 1990s, and Germany may not be immune from similar pressures. Indeed, compared to these cases the United States has a much less severe fiscal crisis. It is not clear, therefore, that this perspective offers much help in identifying the causes of the *specific* problem facing the United States, namely its productivity slowdown relative to its leading 'rivals'.

Profit squeeze 3 – US–world economy identity shift

The final profit squeeze perspective draws attention to the peculiar relationship of the US economy to the world economy since the Second World War. The US government was largely responsible for rebuilding the world economy after the war. It did so by sponsoring a liberal order in which its military expenditures would provide protection for increased international investment and trade. These would, it was thought, redound to domestic American advantage. The logic behind this lay in the presumed transcendental identity between the American and world economies. Over the years, however, a process of internationalization that first brought growth to the US economy has now brought about a challenge to it. The overseas investments of American business have increasingly substituted foreign production for domestic exports from the US. Manufacturing employment and worker incomes have suffered in consequence. Less capital is available for domestic investment and, through taxes, to finance the military and welfare budgets.

This perspective is alert to the irony that 'US' business has held up its share of global production very effectively, but that it has done so through investing outside the United States (Lipsey and Kravis 1987). Whose 'hegemony' is declining? becomes the operative question – that of US-owned business or the US territorial economy? This represents a quite different perspective on the issue of hegemonic decline than any of the other theses by drawing attention to the particularity of US hegemony itself. This perspective is examined in more detail in a later section of the chapter.

National spending 1 – crowding out

The final two perspectives on relative decline both focus on national spending. The crowding out thesis is, with the imperial overstretch thesis, perhaps the most popular perspective of them all. The basic idea is that government spending has increased at a faster rate than has the rate of national economic growth (Figure 5.2). This has accelerated growth of the federal

111

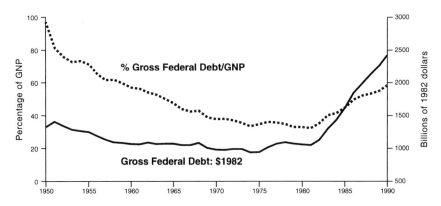

Figure 5.2 Gross federal debt as a percentage of US GNP, in 1982 dollars, 1950–90
Source: Economic Report of the President 1991

deficit which in turn has forced up interest rates and made capital investment more expensive. To finance its deficits government 'crowds out' other potential users of the capital it requires. This problem is deepened by low rates of household saving and high levels of personal consumption. Although there is a capital shortage in the US, this has been masked by America's ability to attract foreign funds through increasing the value of the dollar, the main currency of world commerce.

To pay for what had been borrowed, especially during the 1980s when the federal deficit exploded as a direct result of the policies of the Reagan administration, the US would have had (in 1987 terms) to give up 5 per cent of national income by 1990 through swinging the trade balance from deficit to surplus (Friedman 1988, 53). By the early 1990s the picture was even bleaker (add another 1 per cent). A resolution of the US budget crisis would require both a decline in the net standard of living in the US and the agreement and action of the country's main trading partners.

This is obviously an important problem for the US and world economies. There seems little doubt that the amazing build-up of government debt in the US has contributed to a declining rate of productivity and an increased debt-load for future generations. But rather than simply nefarious government or mistaken economic ideology, the 'bankrupting of America' (Calleo 1992) relates to the changed position of the US in the world economy. Americans consume at levels unrelated to their productivity. Some of this consumption is of military goods to keep the world safe for American (and other) investment. Some of it is of increased expenditures on inefficiently produced public goods like healthcare. But much of it goes to maintain a way of life that came with the early post-war American dominance of the world economy. If the dominance has ended, the way of life continues and the 'day of reckoning' (Friedman 1988) fast approaches.

112

National spending 2 – investment to consumption shift

The final perspective, the investment to consumption shift, sees government spending changing the incentives within the economy to favour income redistribution and consumption at the expense of investment. 'Democracy' encourages citizens to demand ever-more goods and services from government. As the public sector grows and taxes increase, private capital investment goes down. At the same time the services the government provides discourage private saving for the future. This also reduces capital available for investment. The net effect is declining investment and productivity.

The problem with this view is that the United States has much lower levels of public spending on services than do most leading industrial countries. Only Japan spends less of its GDP on public services in percentage terms. Calleo (1992) also points out that Americans see far less of what they pay for in terms of 'civil' public goods than do Europeans. Most healthcare is privately financed and local property taxes finance the lion's share of primary and secondary education. The federal income tax goes largely to subsidize special interests (farmers, homeowners, etc.), to fund the military and military industries, and to finance the federal deficit. If anything, higher public spending, especially on infrastructure and education, and if paid for through taxes rather than borrowing, would help raise national productivity and not diminish it (Calleo 1992).

By way of conclusion, those perspectives which focus entirely on national government spending as the main source of American relative decline are mistaken and miss its most important cause in the relationship between the US and the world economies. The macrohistorical approaches are unsatisfactory because they also miss what is particular and peculiar about the situation in which the United States now finds itself. This is not to say that several of these perspectives do not advance very useful arguments that need to be incorporated into a more thorough account. In particular, the crowding out and leading sector decline perspectives identify factors that are of fundamental importance in connecting the US global position to declines in domestic rates of investment and productivity. But these only make much sense in the context of the changed relationship between the US and the world economy. We return to this theme in the final section of the chapter.

CONTINUING AMERICAN LEADERSHIP

The arguments of 'declinists' have not been without challenge from those who see the United States continuing as the 'system leader' well into the future. The collapse of the Soviet Union and the failure of other potential hegemons, such as Japan and the European Community, to become leaders on 'global problems', have encouraged a view of the United States as a singular superpower untroubled by a serious challenge to its hegemonic status.

Responses to 'declinism' have taken three forms. A first response argues that there has been no decline in the position of the US (e.g. Strange 1987; 1990). According to this construction, proponents of 'relative' decline have missed the continuing absolute power of the United States and focused excessively on the narrow issues of investment and productivity. A second response agrees that there is evidence of decline but lays this at the door of either mistaken government policies or a failure to adopt correct ones (e.g. Nau 1990). A third, and for us more interesting response, suggests that the United States has acquired a larger degree of cultural hegemony at a time when the increased interdependence of the international political economy has given unprecedented importance to co-optive or 'soft' power (e.g. Nye 1990a). From this perspective it is precisely the changing nature of world power that will reinforce the role of the US as the global hegemon.

No decline

Direct rebuttals of the thesis of relative decline emphasize the continuing absolute presence of the United States in the world economy and its importance in the world monetary system. The size of the American market for goods and investment is its most important attribute. The growth of other economies, especially Japan and the other East Asian economies, has been based to a considerable extent on sales to the United States. The growth of imports has undoubtedly hurt some sectors of the US economy, but it has also made other economies more dependent on the US. Increased openness has also stimulated American producers to increase their involvement with the world economy. In the late 1980s the US became the world's most expansive exporter of manufactured goods (Figure 5.3). By world standards US productivity is still impressive and even growth in productivity is not as bad as portrayed by declinists. Growth in business investment has lagged, but it has also stagnated or declined in such countries as Germany and Japan. Now that they have 'caught up' with the US economy they face many of the same problems of aging factories and expensive labour forces that the US faced before them. During the 1980s the United States consistently accounted for about 40 per cent of the aggregate GDP of the OECD (the main industrial countries). This was a 10 per cent decline from the 1960s, but the American economy was still two-and-a-half times the size of the second biggest economy (in conventional exchange-rate terms), that of Japan (Webb 1991, 341).

The incredible power of the United States is especially apparent in relation to global finance. Ironically, the US federal deficit is a monument to the structural power that the US wields through the dollar in drawing on other people's savings to finance its overspending (Strange 1990). A country like Japan has little choice but to go along with American policies for fear of upsetting the dollar exchange rate. 'A devalued dollar would mean either

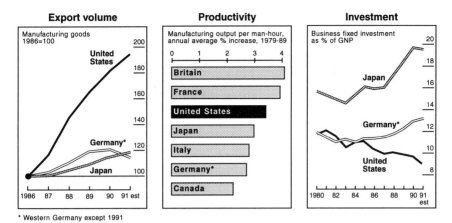

Figure 5.3 US export volume (1986–91), relative productivity gains (1979–89) and business fixed investment (1980–91)
Source: US Department of Commerce 1991

increased competition for Japanese industries in world markets or preventive increases in interest rates with unavoidable repercussions domestically affecting the stability of the national financial system' (Strange 1990, 270). Since the abrogation of the Bretton Woods Agreement in 1971 US governments have consistently attempted to resolve their economic problems by manipulating the value of the dollar against other currencies to gain trading and investment advantages (Parboni 1988). From this perspective American monetary power has actually increased because the new global financial system that replaced Bretton Woods gives greater scope for exchange-rate manipulations. Other countries have had to go along with American policies and desires. The international use of the dollar, like the use of English as a global lingua franca, will not be easily overcome.

The emphasis on absolute as opposed to relative power has much to commend it. After all, in its conventional usage, the term 'hegemony' is about absolute advantages. But the argument for the 'myth' of hegemonic decline fails to note the negative impacts of the changed context in which the US territorial economy must now operate. Above all, the US economy, in the recent past famously removed from the shocks and tremors of the world economy, is increasingly internationalized and subject to influences over which US governments can have little control. Examples include the oil price shocks of the 1970s and the penetration of imports in sectors where US producers were previously unchallenged (e.g. chemicals, electronics, aerospace).

Such changes have had a profound effect on American popular consciousness. The burst of growth and prosperity after 1945 had consequences that were unprecedented anywhere in the world. They created a large pool

115

of well-paid, low-skilled jobs, which in turn lifted worker incomes to undreamt of levels. But this Golden Age did not last long (Marglin and Schor 1990). An economy that had been virtually self-contained for 20 years after 1945 suddenly found itself faced with foreign competition at home. The fall-out in terms of peoples' lost faith in upward mobility and perpetual improvements in living standards has been profound. The international economy is seen by many ordinary Americans as bringing decline rather than furthering a continuation of American hegemony (Newman 1993). Data on the net upward redistribution of income and the shrinking of the middle class in the 1980s suggest that there is something to be explained (e.g. Krugman 1992). Whether it is the direct product of enhanced international competition is another matter (Krugman 1994).

The global position of the dollar is also in question. The argument about the structural power of the dollar depends upon the dollar continuing as the dominant metric of world trade and investment transactions. In 1976 80 per cent of world foreign exchange reserves were held in US dollars. By 1990 this figure had shrunk to 51 per cent, with the D-mark, yen, and a handful of other currencies accounting for most of the remaining 49 per cent of world exchange reserves. As foreign governments move out of dollars and into other currencies, American leverage over the world financial system will surely diminish. They are moving out of dollars precisely because they want to avoid the dollar seigniorage that the US has imposed on them over the past twenty years.

Failed policies

A failure of US government policies is identified by some other commentators as the main source of US relative decline. If these policies are replaced, decline can be replaced by renaissance. Some advocates of the crowding out and investment to consumption perspectives on relative decline share this diagnosis.

Nau (1990) has presented the most coherent case for a restoration of the Golden Age of hegemony. Nau looks to the pre-1967 combination of US domestic and international economic policies for his inspiration. A series of decisions were made at around this time that reflected a 'deep Keynesian consensus' that was fundamentally antithetical to the dollar's role in the Bretton Woods system. However, the 'embedded (institutionalized) liberalism' of the years from 1947 to 1967 – disciplined fiscal and monetary policies, flexible markets and free trade – can be restored so as to re-establish American leadership. The early years of the first Reagan administration (1981–4), Nau claims, showed the kind of decisive action in domestic and international economic policy that is needed to restore American leadership. With the demise of the Soviet Union and a spreading commitment to the liberalization of national economies the time is now ripe for a restoration of the old model; for a return to true and timeless American values.

There are three principal problems with this thesis, even as it is outlined by Nau. First, the 'model' which Nau wishes to resurrect was never a model in the first place. It was a set of practices that evolved into a set of semi-formal arrangements. These never involved a full commitment to free trade. Monetary discipline was enforced largely at the expense of countries other than the United States. Indeed, world trade and global investment flows have expanded much more rapidly since 1967 than they did previously because there are relatively fewer barriers to trade and investment than there were before. At the same time the prospect of renewed US monetary hegemony has become less likely because other countries are now wary of American attempts to manipulate the global financial system to US advantage.

Second, was it poor choices reflecting a mistaken economic ideology that led to US disengagement from the 'Bretton Woods system', or was it a recognition of changed circumstances? Calleo (1982; 1992) suggests that a series of shocks brought that system down. Most obviously, there was the shock of the Vietnam War. Because this war was financed without raising taxes it increased domestic inflation. This brought calls for revaluing the dollar as a means of managing the economy. There were also significant increases in basic commodity prices and increased pressures from internationally-orientated businesses to remove capital controls and shift to flexible exchange rates. Finally, the US government had to do something about the chronic balance of payments deficits that were facing the country in the late 1960s and early 1970s (Calleo 1990). 'Choosing' a new economic ideology had little or nothing to do with it.

Third, and finally, the Reagan policies of the early 1980s are exactly those pointed to by crowding out theorists as precipitating a deepening of the American fiscal nightmare (Friedman 1988, 1992; Calleo 1992). Moreover, as Nau admits, the Reagan administration's reliance on international markets rather than institutional coordination for its international macroeconomic policy had no resemblance whatsoever to the Bretton Woods system. The relevance of the Reagan policies to Nau's thesis about the restoration of Bretton Woods is not readily apparent.

US cultural hegemony

The third and final perspective claims a persisting basis for American hegemony in the changing nature of world power. To Nye (1990a; 1990b) a 'transformation of power' is under way. 'In an age of information-based economies and transnational interdependence, power is becoming less transferable, less tangible, and less coercive' (Nye 1990b, 183). In this context, 'the universalism of a country's culture and its ability to establish a set of favorable rules and institutions that govern areas of international activity are critical sources of power' (182). Although economic and military power remain important, it is no longer the power to command so much as the power to co-opt that will determine which state is hegemonic.

This perspective has much to commend it. Nye's concept of hegemony is much richer than that of most declinists and of the other 'renewalists'. What is less clear is the extent to which American co-optive power is new. We showed in Chapter 2 that American hegemony rested from the start upon a mix of consensus and coercion. The power of the United States never was purely military–economic in nature. More importantly, however, to what extent does American hegemony by means of co-optive power redound to the advantage of the US government and the US territorial economy? American culture has become so 'universalized', to use Nye's term, through films, advertising and other forms of cultural production, that it has lost its specific associations with the United States as a territorial entity. The 'consumer culture' of Western civilization has definitely spread far and wide. But contradictions abound. Boys and girls in Baghdad picking their way through the rubble from a Cruise missile attack can wear jeans and talk in English about 'international' pop musicians yet still hate the US government. The products of American society, such as books, films and videos, have global influence but are increasingly not manufactured in the United States. The connections between American culture, government and economy are increasingly tenuous. Which America is 'bound to lead'?

AMERICAN HEGEMONY VERSUS THE US TERRITORIAL ECONOMY

The conventional understanding of hegemony implies not only a territorial agent but also an 'imperial dividend'; a return to the territorial economy of the state exercising hegemony. Yet, since the early 1970s a case can be made for the increasing divergence between a continuing American hegemony within the international system, on the one hand, and the performance of the US territorial economy, on the other.

Bergsten (1982, 13) identified the critical issues as follows:

> The United States has simultaneously become much more dependent on the world economy and much less able to dictate the course of international events. The global economic environment is more critical for the United States and is less susceptible to its influence.

Capability and autonomy declined in tandem. For example, Rupert and Rapkin (1985) have shown that US shares of global resources and capabilities (measured by the ratio of US GNP or GDP to global totals) have declined in a more or less linear fashion since 1950, with an upturn in the early 1980s. At the same time, the interdependence or susceptibility to 'external shocks' of the US economy (the ratios of US trade and exports to GNP) increased dramatically after 1968. They borrow Bergsten's phrase 'scissors effect' to describe the joint consequences of a declining capability share and increased interdependence.

118

With respect to *capability*, the main problem appears to lie in the decline of GNP potential or the total capacity to produce. This is partly a function of a changing labour force (more dependants, lower skills), but it is more especially a product of declining national productivity. In 1990 this was indexed at 2.5 compared to 4.3 in 1965. As a consequence, annual real growth in GNP slumped from 4.3 per cent in the 1940s through 3.2 per cent in the 1950s, 2.9 per cent in the 1960s, 2.8 per cent in the 1970s to 2.8 per cent in the 1980s. One leading economic forecaster sees US GNP growth heading below 2.0 per cent per annum in the 1990s (Straszheim 1991).

Recent trends also underscore the decline in US relative *autonomy*. The federal government's deficits in the 1980s made the US the world's largest debtor; foreign sources have provided a large part of the credit needed to cover them. More generally, the US is not exempt from the increased interdependence and velocity of the world financial system. Capital has become much more mobile in time and space. Before 1973 there were significant barriers to the international movement of investment and portfolio capital. Many of these barriers have been reduced or removed. The pace of transactions has increased at parallel rates. Before 1973, currency exchange rates changed once every four years on average, interest rates moved twice a year, and companies made price and investment decisions no more than once or twice a year. By 1990 exchange rates were changing several times a day, interest rates moved weekly, and investment decisions were down to once a month or quarterly.

These shifts in capability and autonomy have had important consequences for the economic welfare of Americans. Median household incomes peaked in 1974. Since then the principal feature of American incomes has not been their growth, but their redistribution from poor and middle-class households to richer households. The pace of this movement was especially marked in the mid-1980s with the decline in the number of well-paying manufacturing jobs, an increase in part-time and temporary employment (this avoids the necessity of employers paying health and pension benefits) and tax cuts that disproportionately benefited the affluent. An associated trend has been the re-emergence of inter-regional income disparities and the deepening of central city–suburban income inequalities (Phillips 1990). A resurgence of protectionism and growing hostility to liberal internationalism in the regions most affected by job loss and economic decline (in particular, the old Northeast–Midwest Manufacturing Belt) is one not surprising result.

But how did the US territorial economy get to this point? In Chapter 2 it was suggested that the combination of big business, growing government and overseas business expansion that had created an identity between the US and world economies in the 1950s and 1960s – what was good for one was good for the other – no longer works for the US. As first cause and then consequence of the economic slowdown within the United States, American multinational corporations found overseas investment more profitable than investment in their homeland. Not only did this increase unemployment in the manufacturing sector, but it also reduced the availability of capital for re-tooling factories and investing in product innovation in the United States (Agnew 1987b). The cost of the Vietnam War added to the long-term trend of low rates of industrial investment, balance of payments disequilibria, inflation and political dissent. The 'Cold War consensus' that had prevailed among political élites for the previous two decades collapsed over the conduct of the war.

The Nixon administration was the first post-Second World War American government that had to deal with such a massive crisis. Examining its reaction and the reactions of subsequent administrations provides a good perspective on the increasing divergence between American hegemony and the US territorial economy.

The Nixon administration responded in three ways to the crisis of the early 1970s. First, through cajoling and coercion it encouraged Japan and the Western European countries to revalue their currencies against the dollar. In the short run this made US exports more competitive and imports less so. But its long-term effects were the demise of a system of fairly stable exchange rates (Bretton Woods) and a major boost to a global financial system largely out of the control of any single government (Strange 1986). Second, in terms of strategic nuclear weapons, the United States and the Soviet Union were almost at parity by the early 1970s. During the years that Brezhnev was the Soviet leader (1964–82), emphasis was put on achieving equality in strategic weapon systems with the United States. This did not increase Soviet security nor did it improve the generally low quality of Soviet forces as a whole (Edmonds 1983; Cockburn 1983). But it did increase American perceptions of the Soviet Union as a threat to the territorial United States itself. In this context, the rift between the Soviet Union and China made it possible for the US government both to accept the Soviet Union as a military equal and to use China as a counterweight if Soviet demands were perceived as excessive. This was the essence of the Nixon–Kissinger policy of *détente* with the Soviet Union. But it also carried potential trade and investment benefits if the Second World as a whole could be opened up. Third, as a consequence of the disastrous American involvement in Vietnam, the Nixon administration moved to substitute the (enhanced) arming and support of 'regional surrogates', such as the Shah's Iran, for direct American military intervention. This brought both the advantage of increased exports of military goods and the promise of 'no more Vietnams' in which Americans would die in America's wars.

Whatever the impression given by President Nixon or his foreign policy adviser Henry Kissinger, these policies were responses to the contemporary situation rather than the product of design. The period 1967–74 marked a watershed in America's relations with the rest of the world. An adminis-tration that spoke openly, if ambiguously, of geopolitics, was the first one since 1945 so constrained domestically (by dissent and conspiratorial responses to it: as in the Watergate affair), and internationally (by an increasingly hostile economic environment) that it could not effectively practise it (Schell 1989).

By the mid-1970s the Nixon approach was increasingly in question. Its heavy emphasis on US diplomatic activity and unilateral economic action led to considerable unease among the most 'internationalized' sectors of American business (Gill 1990). How to turn this unease into policy was another matter. In his first two years in office, President Carter tried policies which would make the US less dependent on foreign sources of oil, would stabilize arms competition with the Soviet Union, and would encourage a greater degree of respect for human rights among regimes allied to the United States. These policies foundered, however, because of three developments. The first was the influence of right-wing groups inside (in the person of the National Security Advisor Brzezinski) and outside the Carter administration that identified the Nixon policy of *détente* as nothing short of an appeasement of the Soviet Union. Although these groups (such as the Committee on the Present Danger) differed in their proposals, all were agreed that the US needed to reassert itself militarily (Dalby 1990).

A second development related to the international debt crisis which broke in the early 1980s on the back of excessive commercial bank lending to some middle-income countries. When Paul Volcker raised interest rates in America in 1979 the cost of servicing dollar-denominated debts escalated sharply and unexpectedly (Corbridge 1993). Debt servicing became impossible for some countries when oil prices rose again in 1980 and when demand for the borrowers' products went down in the global recession of 1979–82. The US was faced with the spectre of lending bank failures if several of the largest debtors (e.g. Brazil and Mexico) defaulted simultaneously. Some of the chief problems of the US financial system in the 1990s date back to this era. In particular, commercial banks began to lose their central place as sources of capital (from 38 per cent of financial assets in 1975 to 24.5 per cent in 1993) to mutual funds and brokerage firms (Bacon 1993). This was exacerbated by a process of deregulation in the 1980s that favoured savings and loan banks (many of which later defaulted at great government expense) and the stock market. The declining role of commercial banks has made the US financial system more vulnerable to external shocks transmitted through the stock market and less able to supply credit to small firms interested solely in US markets.

A third development was more by way of a collapse. In 1979 both the United States and the Soviet Union were faced with the disintegration of surrogate regimes in, respectively, Iran and Afghanistan. In other regions, too, such as the Horn of Africa and Central America, surrogates proved to be unstable and unreliable (O'Loughlin 1989). Superpower competition could no longer be conducted indirectly through the use of surrogates. The Carter administration was defeated in the 1980 presidential election largely because of its failure to maintain the Shah in power in Iran and by the hostage-taking episode which accompanied the rise to power there of Ayatollah Khomeini.

In the absence of a viable territorial strategy involving a national industrial policy, capital-export controls, and a decline in military spending, the 'only' tried and true solution to the multiple dilemmas open to the US was the revival of militarization. This had worked in the early 1950s to stimulate the economy and discipline dissent; both the later Carter administration and the Reagan administrations of the 1980s chose it again. Increased military spending in the US stimulated investment and employment, at least in those regions where it was concentrated (e.g. California, Texas, Connecticut), and increased military commitments abroad demonstrated an American resolve to reassert its authority in a world that America had created. To President Reagan, it was 'morning again in America'.

The first Reagan administration carried through the remilitarization strategy begun under Carter to a level not seen since the Second World War (Figure 5.4). Reagan's economic goal was to jump-start the US economy with a massive programme of government-sponsored, weapons-based industrialization. The geopolitical goals were to remind Europe and Japan of their reliance on an American military commitment that underwrote their economic development, to counter the Soviet attempt at strategic–military equality with the United States, and to encourage Third World countries friendly to American economic interests.

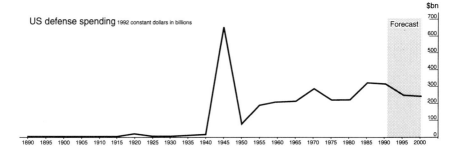

Figure 5.4 US defence spending, fiscal years 1890–2000, in 1992 US$ billions
Source: US Defense Department 1992

In addition to massive military spending the Reagan administration attempted to reverse the American competitive slide in the world economy in three ways: recession-led restructuring, tax cuts and deregulation. In order to bring down inflation, which in 1978–9 was at a post-1945 historic high, the first order of business involved tightening monetary policy. This was done so drastically between 1979 and 1982 that the US experienced its most severe economic downturn since 1937. Attention was then switched to the 'real' economy. Income tax rates, especially for the very rich, were cut dramatically on the assumption that the extra money put into circulation would end up as new productive investment. Finally, deregulatory policies were introduced to allow 'the market' to choose between candidates for growth and candidates for bankruptcy. The main goal was to encourage mergers and acquisitions, but subsidiary policies included reductions in environmental and safety regulations, assaults on unions and the elimination of many rules governing the conduct of financial institutions.

In the short run these policies were such a stunning 'success', particularly in reducing that traditional bane of the middle class, inflation, that 'Reaganomics' produced a landslide re-election victory for President Reagan in 1984 and widespread imitation abroad. The US political economy's 'peculiar qualities' by the standards of the industrial world as a whole – lower public spending on services, low levels of worker unionization, etc. (Rose 1989; Johnston 1993) – became international 'standards of excellence' for economic performance (Davis 1985).

But the long-term consequences for the US territorial economy have been nothing short of disastrous. In the first place, the tax cuts did not produce a bonanza of capital investment in new plant and equipment. The money went largely into consumption (often of imports), into a frenzy of mergers and acquisitions among large firms and, in a world economy with minimal capital controls, into overseas investment (Friedman 1988, 233–70).

As a second consequence, the federal budget deficit exploded from $59.6 billion in 1980 to $202.8 billion in 1985. In lieu of the taxes foregone after 1981, the increases in military spending had to be financed by borrowing. Cuts on the expenditure side of the budget were insufficient to make up the difference. Because the US savings rate was so low, most of the borrowing had to be done overseas (Friedman 1988, 209–32). When added to the trade deficit this produced an explosion in the US current account balance (Figure 5.5). In the

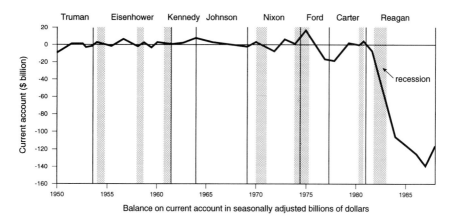

Figure 5.5 US current account, 1950–88
Source: US Department of Commerce 1990

123

past, even when the US ran a trade deficit, or when its basic balance of payments was in deficit (because of military costs abroad, travel expenditures and remittances abroad from people working in the US), the current account (trade plus overseas earnings plus royalties and fees) had been in surplus. This was no longer the case after 1981.

Whether this matters as much as it did in the past is open to question. According to one school of thought, as long as financing can be found the 'nationality' of its origin does not matter; with variable exchange rates and few restrictions on capital mobility there is no longer the old problem of running down foreign-exchange reserves. However, from another perspective, trade and current account deficits do give a reasonable picture of the relative state of a particular national economy and they are still used by governments and private investors as indicators of national economic health. They are not simply 'tricks' of the economist's 'trade', even if with globalization they have lost some of their macroeconomic significance.

A third consequence was that Reaganomics had very specific geographical effects within the United States. Generally, the two coasts benefited at the expense of the mid-portion of the country. There was a large redistribution of income in favour of already wealthier groups and regions (Phillips 1990). Military spending pumped huge sums into southern California, Texas, New England and Washington State, just as it had in the 1950s (Trubovitz and Roberts 1991). Other regions, particularly the Midwest, received relatively little. Previously well-paying manufacturing jobs in these regions had disappeared or now paid lower wages, largely because of import penetration (Abowd and Freeman 1990; Markusen et al. 1991). Inter-regional income inequalities, which had been declining throughout the post-1945 period, began to increase. As a result of both import penetration and tax and military-spending decisions, the US was more socially and geographically polarized in 1990 than it had been at any time since the 1930s (Figure 5.6). Only the dramatic cutbacks in defence spending in 1992 have created something of a (negative!) counter-trend. As a result of the end of the Cold War, and the need to reduce the federal deficit, regional economies like southern California and parts of Texas, have been devastated. This shows the extent to which the prosperity of these regions was built on the militarization of the US economy during the Cold War.

Of course, the current account and federal budget deficits could be justified if they had led to a fundamental upward shift in investment and productivity. But the available evidence suggests that they have not. Average annual rates of growth of GNP in the 1980s as a whole were no higher than those of the 1970s (Friedman 1988, 187–208). The deficits also pose a danger to the world economy. To finance its current debt the US government draws in resources from abroad that could surely be used more productively elsewhere. To deal with the underlying imbalance in trade and investment the US net position in world trade would also have to change by about +6 per cent of total world

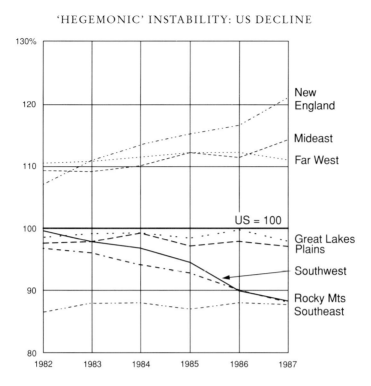

Figure 5.6 Regional per capita incomes as a percentage of US average, 1982–7
Source: US Bureau of the Census 1989

exports. To do this, Japan and Germany, among other countries, will have to run trade deficits with the United States of a magnitude equivalent to the necessary US surplus (Corbridge and Agnew 1991). With the collapse of the Soviet threat that has always disciplined Western Europe and Japan into accepting American demands, there might be increasing doubt that these other countries will act so magnanimously. At the same time, the constraint of unprecedented global interdependence does provide a new bond between the US and these other countries. They may not be free to choose.

After twenty years of tinkering at the interface between the US economy and the world economy, US government policies have deepened rather than resolved the country's underlying dilemmas. Within the United States the seeming success of American hegemony (*vis-à-vis* the former USSR), and its continuing power in the form of transnational liberalism are increasingly challenged by those who see them as hollow victories. From one point of view, what really matters is whether the US territorial economy can provide rising living standards for its population through enhancing the strength and competitiveness of the economic activities located within its borders irrespective of the nationality of ownership (e.g. Reich 1991a). In this perspective,

Money is unpatriotic; these days, investment dollars are speeding to wherever on earth they can get the highest return. People, however, are relatively immobile, and they belong to societies with particular cultures and histories and hopes. It is up to governments to represent people, to respond to their needs and fulfill their hopes – not to represent global money.

(Reich 1991b, 53)

From another point of view what really matters is that America's status as Number One is restored by moving from the geopolitical conflict of the Cold War to 'geo-economic' conflict with states that do not trade 'fairly' (e.g. Luttwak 1990; Huntington 1993a). Japan is usually singled out for special attention in this regard (O'Tuathail 1993b). Japan's lack of incoming investment relative to outgoing investment is identified as a deviation from an American model that is held up as the contemporary paragon of economic virtue (Figure 5.7). The chief issue here involves the declinists' standard concern; namely, the relative aggregate economic gain of Japan (or another economic challenger) over the United States, and US downward economic mobility within the world economy (e.g. Mastanduno 1991). Except for some interest in the increased dependence of the US government on foreign suppliers of weapon components (see Moran 1990), there is little or no attention given to the 'deterritorialization' of the world economy to which data on foreign direct investment such as those in Figure 5.7 directly attest.

What is most problematic about both schools of 'national competitiveness', however, is their failure to note that the condition of the US territorial economy even as it 'deterritorializes' is still governed largely by domestic considerations. The US domestic rate of productivity growth is largely the result of decisions concerning finance and investment in technology, training, location of facilities and infrastructure made by firms and governments within the United States. 'Foreigners' are not the main threat to US living standards, decisions made by Americans are (see Krugman 1994; Krugman and Lawrence 1994). Of particular importance, US government decisions in the 1980s undermined domestic productivity growth.

The lack of centralized 'decisional power' within the Madisonian system of US government (Friedberg 1992), together with the entrenched power of transnational liberalism with its supports in the internationalist business establishment, mean that probably neither of the 'new' perspectives on national competitiveness is likely to prevail. This is an even more likely outcome (or non-outcome) when one bears in mind that national stagnation or low growth itself fosters conflict over relative group or regional gains. What is the 'national interest' in the contemporary global context? It will be difficult to re-establish the kind of national consensus that characterized the 'internationalist' coalition in the US from the 1940s to the 1960s. The old identity between the US and the world economies on which this rested has gone.

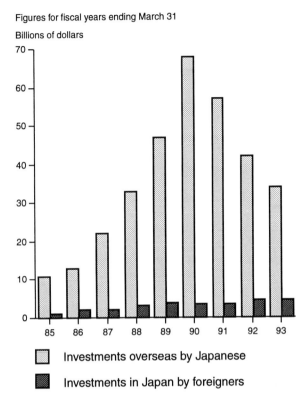

Figures for fiscal years ending March 31

Billions of dollars

▫ Investments overseas by Japanese

◼ Investments in Japan by foreigners

Figure 5.7 Japan's investment imbalance, fiscal years 1985–93
Source: OECD 1993

CONCLUSION

The previous section should not be read as indicating an imminent political–economic apocalypse for the United States. That is not on the cards. Within the world of territorial states the United States is still *primus inter pares* and it will remain so for the foreseeable future. US hegemony is nowhere more apparent today than in the military arena, where it was demonstrated in the Gulf War (albeit with financial support from Europe and Japan and under the cover of the UN), and where it has been reinforced by the break-up of the Soviet Union. The dollar also remains the world's leading reserve currency, and Wall Street continues to serve as one of the most important nodal points in a world economy increasingly made up of widely diffused groups linked together by high-speed information flows. It also bears repeating that the GDP of the US territorial economy is still greater than that of Germany and Japan combined, and that the US is at least as well-placed as its rivals to capture those future waves of technological innovation that will

127

emerge from a less geographically constrained global market-place. The United States economy is bound to decline further in relative terms simply because other economies have improved their performance, but there is no reason to assume that the pace of decline in the US will be greater than that in Europe and Japan. If the United States can come to terms with the burdens of the Reagan years – the twin deficits – it may yet surprise its competitors in the advanced industrial world. Certainly, there is no singular hegemon waiting to take its place; no hegemonic succession waiting to happen (see Chapter 6).

These points made, we do still believe that the relative decline of the US territorial economy has implications for the position of the United States as a 'hegemonic power'. This is true in two main respects. Consider, again, the military arena. It is true that the size of the US nuclear arsenal, and the sophistication of its weapons technologies, make it unrivalled as a military superpower. At the same time, the American nuclear umbrella has also afforded protection to America's allies in NATO, including Germany and Japan. Precisely because the nuclear balance of terror has kept the 'peace' so well, it has not been necessary for these countries to invest heavily in the conventional hardwares of 'defence'. If there is now a peace dividend for the US to reap, it is not unreasonable to suggest that Germany and Japan have been reaping one for over forty years. The US also had to come to terms with the threat of nuclear proliferation – especially since the break-up of the USSR. The Ukraine was for a while the third largest nuclear power in the world. This must inevitably place new strains on the US in its role as 'global policeman'; either that, or it will lend encouragement to those groups within the United States who are demanding that Europe and Japan pay more visibly for their own defence. But herein lies another catch. Germany and Japan may not yet be military superpowers, but they have the economic might to assume this position. US strategy during the Cold War was as much about restraining this possibility by integrating these fallen enemies into the American 'free world' economy as it was about containing the expansion of its alter ego, the Soviet Union (Cumings 1992).

The Cold War also provided an overarching geopolitical discourse to guide US military policy. Without the linkage to a global geopolitical contest it is hard to identify priorities and justify particular decisions. With global dualism removed, defining the 'national interest' becomes a case-by-case endeavour. This makes domestic politics in the US more divisive; there is less agreement possible about 'where' the highest stakes lie. It also makes defence planning and production much more difficult. What kinds of wars and what kinds of weapons are the wave of the future? The disarray in US government policies towards Somalia, Bosnia and Haiti in the early 1990s is symptomatic of the impact of both domestic political divisions and the difficulty of planning for a post-Cold War world.

And then there is the matter of economic hegemony. The US territorial economy is still the largest territorial economy in the world, but in a world system held together by information flows and new communications technologies it is not clear that territorial states are any longer the world's dominant economic actors. Even the US government has difficulty today in maintaining a target exchange rate for the dollar, or in holding down interest rates below those demanded by international investors. In a very real sense, the US territorial economy is now hostage to an international economic system that American hegemony did so much to unleash after 1945. Many powers that once accrued to countries have now leaked to the 'markets' and to groups of 'market-makers'. The United States is still a 'hegemonic power', but the nature of the hegemony it exercises – in form and in degree – is quite different to the hegemony it exercised in the 1950s and 1960s. The novel context in which the US must now operate is the subject of Chapter 7. The question of which other states could possibly displace the US, given a continuing process of hegemonic succession, is addressed in the next chapter.

6

'HEGEMONIC' PRETENDERS

For purposes of exposition this chapter is based on the premise common to hegemonic succession views that eventually a new hegemonic state (a hegemon) arises to replace the old one. In neo-realist terms we are presently in a 'unipolar moment' after the bipolarity of the Cold War, with the United States now supreme. But there are a number of candidates for international 'primacy' available to replace the United States as 'Number One', unless it acts decisively to prevent this. In the interim there is likely to be an emerging system of 'multipolarity', which in the dominant view is inherently unstable and cannot but in time produce competition for the mantle of hegemon. Of course, to repeat earlier admonitions, this is not the view of the nature of 'hegemony' espoused in this book or in the next chapter. But the political and intellectual prominence of neo-realist views requires that their claims be taken seriously.

The main purpose of this chapter is to examine the case for actual hegemonic succession after American hegemony. The first section addresses the argument for the emergence of new Great Powers: that unipolar moments cause geopolitical 'backlashes' which lead to multipolarity and struggles for hegemonic succession. A second section examines the collapse of the Soviet Union as the ultimate stimulus to the emergence of US unipolarity and considers the prospects for the emergence of Russia as a Great Power. The next four sections offer short reviews of the prospects for China, Japan and Germany as 'hegemonic' pretenders. A final section examines the evidence for the emergence of competitive 'trading blocs', the European Community (European Union), the North American Free Trade Area, and East Asia in particular, as a replacement for the American-sponsored transnational liberalism presently in ascendency.

THE PURSUIT OF PRIMACY

The basic assumption of much of the literature on hegemonic succession is that other states 'balance' against a hegemon when it is no longer checked by a serious rival (e.g. Gilpin 1981; Modelski 1987; Kennedy 1987). In particular, with the collapse of the Soviet Union as a military rival to the United States,

Japan and Germany are no longer constrained by fear of the Soviet Union to remain within a security and economic system led by the United States. China, another emerging economic 'superpower', is also frequently mentioned as another potential Great Power. These states, and, perhaps, a revitalized Russia, are expected to emerge as the main candidates for hegemonic status. Three questions arise: (1) What makes a Great Power (and potential hegemon)? (2) Is the pursuit of primacy still possible? (3) Are these particular Great Powers candidates for hegemonic status? The chapter is concerned with trying to answer these three questions. This section addresses the first two.

The emergence of Great Powers typically is seen as a structural phenomenon, resulting from the combination of two phenomena: uneven rates of economic growth and international anarchy. These are ideas redolent of the naturalized geopolitical discourse originating between 1875 and 1945. But they are central to claims about hegemonic succession. The first of the two phenomena refers to the idea that the power of states grows at different rates, largely as a result of differences in rates of economic growth. In this construction, as some states gain power others are losing it. Power is always being gained or lost relative to others, rather than possessed or not possessed in an absolute sense. It is as if there is a global 'pool' of power of fixed volume which when drawn on by one state at an increasing rate diminishes that available to others. To Gilpin (1981, 13) 'the differential growth of power of various states in the system causes a fundamental redistribution of power in the system.' The outcome, as Kennedy (1987, xxii) asserts, is that, historically, 'relative economic shifts heralded the rise of new Great Powers which one day would have a decisive impact on the military/territorial order.' Inevitably, therefore, the higher rates of economic growth of states other than the US will undermine American hegemony.

Several important phenomenological assumptions are necessary to see this relative shift in rates of economic growth translate into Great Power emergence. First, states must want to advance their relative growth advantage into an increase in status within the state system. Second, rather like stereotypical capitalists accumulating capital, states are not satisfied with small increments of status. They want to be 'top dog'. Third, rising power leads to increased international commitments and military obligations. Uneven rates of economic growth, therefore, have cumulative effects that trigger the rise and fall of Great Powers.

The second phenomenon is international anarchy. The state system is seen as intrinsically competitive. If one state was to give up its ambition to be a Great Power another state would replace it. This is why relative power is tagged as key. If another state increases its relative power then a state must either imitate its success (called 'sameness' by Waltz (1979, 127)) or balance against it by combining with other states. The former is a hegemonic strategy, the latter a balance of power strategy. In either case a unipolar situation is likely to be momentary. Sooner or later, states seeking upward mobility will act to reduce their subordination. One implication of this could be that a declining hegemon such as the United States should recognize this and act to encourage multipolarity. Its geographical isolation, nuclear weapons and massive economic capabilities would guarantee the United States an important balancing role in a multipolar system (e.g. Layne 1993). Another implication, however, would be that the US should act to reassert its preponderance before it is too late. Given the assumptions about the intrinsic nature of inter-state competition, hegemony by another state will probably follow from multipolarity (e.g. Huntington 1993a).

Such assumptions are critical to the prospect – or notion – of one of the hegemonic pretenders displacing the United States as global hegemon. To what degree do they reflect current practices? Very little, we would claim; they are throwbacks to another era. There are five specific problems with them. First, as Jervis (1993) argues, the concern with relative power reflects a world in which the threat of war between Great Powers requires constant attention to preparedness and relative advantage. 'In the current era, we need to begin by asking whether nuclear weapons, especially the possession of second-strike capability, have rendered this approach obsolete' (Jervis 1993, 55). What matters in the nuclear era, and, hence, why there is so much concern with the dangers of nuclear proliferation, is not the relative size of weapons stockpiles but the absolute ability to inflict deadly damage. Though worried about the possibility of complacency in the face of the unlikelihood of war breaking out among Great Powers in these circumstances, Jervis (1993, 56–7) is alert to the psychological danger of the focus on relative power: 'an American policy that stresses the possibility of great power wars and the concomitant continued importance of relative gains might turn out to be a self-fulfilling prophecy that will make such wars more likely.'

Certainly, the focus on relative gains appears to characterize much of the debate over the economic challenge posed to the United States by Japan. For example, in a frequently told story, the Harvard political economist Robert Reich (later Secretary of Labor in the Clinton Administration) asked a set of groups of students, investment bankers, professional economists, citizens of the Boston area, and senior State Department officials this question: for the United States which of the two following scenarios is preferable? (1) one in which the US economy grows by 25 per cent over the next ten years, while that of Japan grows by 75 per cent or (2) one in which the US economy grows at 10 per cent while the Japanese economy grows 10.3 per cent. Most people in each group except one chose (2). The economists, thinking quantitatively, unanimously chose (1). The magnitude of the difference in (1) may have pushed some people towards (2). What is clear, however, is that most of the respondents were willing to forego a larger absolute increase in 'their' economic well-being to prevent a larger relative advantage to Japan (Mastanduno 1991, 73).

US government economic policies are also inspired to a certain extent by a conception of relative rather than absolute gains. For example, US policies on satellites and the FSX fighter aircraft in the 1980s reflected a developing discourse about relative gains and losses (Mastanduno 1991; O'Tuathail 1992c). From one point of view, this focus is the outcome of the end of the Cold War and the disappearance of the incentives which that conflict gave to the United States to operate in 'Western' rather than national terms. However, it also fits the self-fulfilling prophecy logic identified by Jervis. Irrespective of the real evidence for the complex interdependence of the US and Japanese economies, American élites and the American public still view the international political economy to a considerable extent in state-territorial terms (also see Chapter 5). Discourse counts.

Second, economic growth and prosperity among the Great Powers have been enhanced since 1945 through the freeing of trade and increased international investment. The focus on relative gains reflects a mercantilist economic ideology in which national economies are equivalent to 'containers' which can only be filled at one another's expense or through territorial expansion. Yet rates of economic growth today are greater when national economies are open rather than closed (e.g. Krueger 1992). Access to international markets is the principal precondition for sustained economic growth. In turn, increased interdependence reduces the incentive for resolving disputes militarily. A state's economic assets are no longer tied entirely to its territory. Foreign ties and stakes are a force for symmetrical interdependence when they are shared by all parties. This reduces the element of conflictuality and increases the return to cooperation.

Of course, the globalization trend could always reverse. The world economy at the end of the nineteenth century was more integrated than it is today in terms of the importance of trade between the leading territorial economies and the openness of national capital markets (Knox and Agnew 1994, 128–9). Economic nationalism could always intervene again as it did before. However, this time around the amount of world trade within firms (about 40 per cent of US trade is intra-firm) and the global ties of firms to multiple locations in different states are such as to make the costs of reversion to autarky much greater than they were in the past.

Symptomatic of the 'internationalization of the state' is the growing concern of firms and states with 'market access'. While states continue to control the rules and structures of their national economies, through formal international agreements and informal but explicit bargains, they also grant access to 'foreign' competitors. Classic free trade and investment policies, therefore, are being extended into the areas of industrial policy and domestic regulation to facilitate cross-border business collaboration (Cowhey and Aronson 1993). This is one of the most important consequences of the growth of world-regional trading blocs, especially the EC, but it is also at the centre of the recent (Uruguay) round of the GATT.

133

Third, American hegemony is not substantively equivalent to other instances of hegemony; it has been institutionalized internationally in a wide range of organizations (e.g. the IMF, GATT, etc.) and has had a widespread cultural influence. Though the 'hegemonic' pretenders could abolish or adapt such institutions and influence, none has the universal political and economic values (and arrogance) that characterize the United States. The demise of the Soviet Union has left the United States as the only leading power with a 'global message': mass consumption, personal liberty, private property, markets and electoral democracy. In a world with massive material deprivation and immense inequality this is an attractive message for both political élites and poor populations.

The most important consequence of American hegemony has been the enmeshment of states in a network of formal and informal regimes that constrain their autonomy (Zacher 1992). One illustration of this is the growth in the number of international organizations, many of which require the adherence of member states to specific rules of conduct. Concerning this expansion in international organizations and their activities, Claude (1988, 117) argues that it 'suggests that statesmen are now more willing to emphasize collective rather than merely unilateral approaches to a wide range of issues'.

Fourth, the number of Great Powers is small and the incentives for cooperation between them now are relatively great. Even at its height in the 1950s and early 1960s American hegemony involved bargaining and compromise with the other leading industrial capitalist countries (Mastanduno 1988; Kahler 1992). Diffusion of power or pluralism, more than multipolarity among singularly 'selfish' states, is now the evolving trend among the Great Powers (e.g. Rosenau 1990, 113). Each state has a stake in seeing the others adhere to common rules governing security, trade and investment (Conybeare 1984; Jervis 1985; Milner 1988). The liberalization of the Japanese economy and the opening of the Japanese political system to changes in the parties of government, for example, are indicative of this growing commonality, notwithstanding continuing differences between the various Great Powers in how they organize their political economies (Rosecrance and Taw 1990). Even increasing the number of participants in multilateral international governance (as in the Law of the Sea and other international negotiations) may not be the barrier to successful cooperation that it was once thought to be (Kahler 1992).

This point is worth emphasizing for what it says about the *social* nature of the international political economy (Caporaso 1992). Conventionally, interdependence between states has been thought of as inherently conflictual, as in the Prisoners' Dilemma simulation so beloved by game theorists. In this perspective a two-party game will not produce cooperation since each party acting as an isolated individual believes it can gain more by not cooperating. Of course, the assumption that states can be viewed as rational and moral agents is a controversial one and social choice theory has all but ruled out the coherence of the concept of 'national interest' (Arrow 1951; Parfit 1984). But if, for the sake of argument, game theoretic reasoning is accepted and pursued beyond its usual limits, some surprising results are produced. For example, because of institutional development and greater interaction, Prisoners' Dilemma 'games' among the potential Great Powers have been increasingly converted into cooperative ones (Snidal 1985; Stein 1991). This is still much less the case for relations between other states and for the asymmetric interdependence that characterizes the links between the industrial capitalist states and the underdeveloped states (Goldgeier and McFaul 1992). But in the context of the idea of hegemonic succession it is relations between Great Powers that count, and the balance there has definitely shifted from conflict toward cooperation, even within the premises of game theory.

Fifth, and finally, there is a growing élite and popular consciousness in Europe and North America of international interdependence, both in general and in relation to certain specific problems or 'issue areas'. Slowly and fitfully a sense of a common global fate has begun to arise. This has been encouraged by the spread of access to information about previously distant places. But it is also the result of the growing sense of linkages and dependencies in a more interdependent world. This does not immediately translate into an instant bonding with distant others, as the more naïve versions of the 'global village' implied. What we have is a common space. Intercontinental nuclear weapons were perhaps the first stimulus to this new more global imagination. More recently, publicity about global ecological problems, such as the thinning of the earth's ozone layer and global warming, and the emergence of international resource conflicts over the use of oil and water and access to forests and arable land, have added considerably to demands for international agreements that recognize a common global destiny (see, e.g. Homer-Dixon et al. 1993; Lowi 1993).

It is the pursuit of primacy itself, therefore, rather than simply the identity of the 'next' hegemon, that is increasingly in question. The presumption of a hierarchy of states produced by competition among putative or original equals makes less and less sense yet still drives much understanding of the international political economy. But, for the sake of argument, accepting its continuing plausibility, what are the qualifications of the main candidates for successor hegemon to the United States?

THE SOVIET COLLAPSE
AND RUSSIA AS A GREAT POWER

The sudden and surprising collapse of the Soviet Union in 1989–91 removed one of the two territorial pillars of the Cold War Geopolitical Order. Not only did the organization which provided both the ideological justification and centralized discipline, the Communist Party, disintegrate, but so too did the territory and the national economy run from Moscow. It was a complete political, territorial and economic collapse. However, several of the successor states, Russia, Ukraine, Belarus and Khazakhstan, have large numbers of nuclear weapons and missiles capable of delivering them. Russia is the world's largest state by land area and physically dominates the Eurasian landmass. It also has large reserves of important industrial raw materials. In mercantilist terms Russia is at least a potential Great Power. As the Soviet Union was commonly acknowledged to have been a 'superpower', the prospects of Russia as the largest remaining component must be assessed in terms of what undermined the old entity. To re-establish itself Russia must be able to come to terms with the problems that plagued it as part of the Soviet Union.

There is considerable controversy about the causes of the Soviet collapse. Many accounts of the Cold War assumed that the Soviet Union was a permanent feature of international relations destined to endure for the foreseeable future. It has been difficult for the field of international relations to cope with a world without the Cold War, never mind explain its demise. This may be because the main emphasis was on the US–Soviet relationship abstracted from the shifting fortunes of the international political economy. The account offered in Chapter 2 offers the necessary elements for an explanation.

The Cold War Geopolitical Order, as the earlier discussion (in Chapter 2) emphasized, had two overlapping but distinct parts: (1) the Cold War relationship proper with the geopolitical division of the world and ideological conflict as its main features and (2) a world economy under American auspices, marked by growing global interdependence through shared institutions and 'regimes' for managing international transactions and expanding world trade and investment. In this context the Soviet Union's capacity to sustain its position as a military superpower was limited by its territorial economy. The United States was much less severely constrained because of its interactional links to the expanding world economy. Stability existed 'within the Cold War system so long as the Soviet Union's economic resources were sufficient to maintain political and military control over its own domain' (Crockatt 1993, 76). This ceased to be the case after the early 1970s, and continued Soviet economic isolation meant lack of access to the rapidly developing financial and technological assets of the world economy. The Soviet Union was no longer a model of modernity even for elements in its own leadership.

In the most thorough comparison to date of the various possible explanations of the Soviet collapse, Deudney and Ikenberry (1991) identify the lack of industrial modernization as the key factor. Soviet rates of economic growth slackened in the 1960s and economic stagnation was the rule through the 1970s and early 1980s. Technological innovation peaked in the late 1960s (Kontorovich 1992). For example, the main aeroplanes of the state airline Aeroflot still flying in the late 1980s were state-of-the-art 1967 and not dissimilar from the American planes of the 1960s long since scrapped or sold for use in Africa or Latin America. The Chernobyl nuclear accident of 26 April 1986, which because of its effects beyond the boundary of the Soviet Union proved impossible to cover up, drew attention to the shoddiness of the Soviet nuclear power industry and the failure of a centralized bureaucracy to protect Soviet citizens from its inferior technology.

All the while, as the economic assets of its adversaries increased, and Soviet economic performance remained sluggish, the Soviet Union was pushed to spend ever-greater proportions of its national resources on its military (Rowen and Wolf 1990). This may well constitute a classic case of the 'imperial overstretch' thesis that Kennedy (1987) wants to extend to the decline of all Great Powers. The autarkic nature of the Soviet economy provides the key variable that explains the success of the thesis in this case.

But relative *economic* decline can be overplayed. By the late 1980s the entire Soviet system was in crisis. The economy now lagged qualitatively rather than quantitatively in what was produced. If it did nothing else the American Strategic Defense Initiative (SDI or 'Star Wars') of 1983 was a reminder of how far behind even in military technology the Soviet Union now was. Soviet élites were well aware of the progress made by the NICs and of the advent of the new informational technologies that promised a total reorganization of industrial production from the nineteenth-century model that prevailed in the Soviet Union. But computers and new information networks threatened the central control exercised by the apparatus of the Communist Party (Judy and Clough 1989). The Party had been in political decline for years. The ideological conflict with China gave it a boost that lasted well into the 1970s. But by the late 1980s it was politically and morally exhausted. Jowitt (1992, 255–6) identifies the following as the four 'big reasons' for the collapse of the Party and its Leninist alternative to capitalism in its various forms: 'Khrushchev's disavowal of class war, Brezhnev's neotraditionalization of Soviet society, the appearance of a revolutionary national citizen class (Solidarity) in Poland, and Gorbachev's relativization of Lenin's absolute Party.' Collapse was a long time in the making.

The nature of the collapse is perhaps as remarkable as the fact of the collapse itself. Prior to the establishment of the Gorbachev government, the chief expectation of those commentators in Europe and the United States who saw economic stagnation in the USSR as a factor threatening the status quo was that it would lead to renewed expansionism and authoritarianism (e.g. Luttwak 1982; Pipes 1984). They were wrong. The Soviet government responded to its crisis with a mixture of internal liberalization and external accommodation. To understand this, Deudney and Ikenberry (1991/2) point to the changed international environment facing the Soviet Union in the late 1980s. Outside influences may have been minimal in directly causing the Soviet internal crisis but they decisively shaped its response to the crisis. The world outside was now more enticing than threatening. Solutions to internal problems could no longer be sought without external contact. The pluralist and pacific features of the external world made making contact less problematic. Above all,

> The inability of the Iron Curtain to isolate the Communist world from the influences of the West and the inability of either side to protect itself from nuclear destruction are as important as the ability of the West to frustrate Soviet territorial aggression. Thus Soviet changes are significantly the consequence of the inability of modern states – even large, socialist states – to exclude influences from beyond their borders.
>
> (Deudney and Ikenberry 1991/2, 115)

The new Russia will have to live in this opening world without the illusion that all transactions can be controlled through territorial mechanisms such as central planning and the Party hierarchy. This places limits on the possibility of restoring a territorial-autarkic economy. And it also puts limits on the possibility of recreating an authoritarian central government. As yet, however, existing identities and boundaries have broken up without any breakthrough to new ones. The Party and the State have been for seventy years the dominant foci for political identity. Socio-economic cleavages were targeted as illegitimate means for organizing political life. Ethnic rather than civic consciousness has been the first fruit of reform, possibly because the republics of the former Soviet Union were organized around officially recognized 'nationalities'.

Creating nationwide interest groups and a politics organized around the divisions of Russian civil society will prove difficult now that ethnic differences have assumed central importance. The main ground for optimism, however, is that the collapse of the Communist Party has removed the chief barrier to the improvisation of new institutions based on social solidarity. Writing before the denouement of 1991, Hosking (1990, 49) noted that beginning in the Brezhnev years:

> A society which had been atomised and traumatised under Stalin was slowly and painfully reconstituting its memory and reknitting the fabric of social solidarity. It turned out that the Soviet people were not afflicted by total amnesia. A dim sense of loss of community was reawakening, together with an intellectual inheritance which had not totally succumbed to the totalitarian rewriting of history. So a certain basis did exist for the revival of autonomous social institutions and eventually of genuine politics.

Two factors are likely to condition the transition from a closed to an open political economy. A first factor concerns the extent to which Russian nationalism draws the country back into itself, or into interventions in adjacent states with large Russian minorities such as the Ukraine, Latvia, Estonia or Khazakhstan over questions of citizenship rights and the disposal of the assets of the former Soviet Union (e.g. the Black Sea Fleet and control over its port of Sevastopol in the Crimea, since 1954 part of the Ukraine).

But here again change is visible. According to some commentators an important feature of the recent past has been growing disillusionment with empire on the part of ordinary Russians. Except on the political fringes, with anti-Semitic and anti-Western groups such as Pamyat, the traditional association of Russian nationalism with imperial consciousness has begun to fade. '[T]here has been a notable shift in recent years away from the former imperial consciousness to a profound disillusionment with empire, accompanied by a widespread conversion to the ideas of liberal Russian nationalism' (Lapidus *et al.* 1992, 9). Less optimistic views, however, have been reinforced by the success of the far-right in the 1993 parliamentary election (for example, see Laqueur 1993). The main figure in this success, Vladimir Zhirinovsky, speaks openly of establishing Russian dominance in Central Asia and the Middle East. Indeed, his geopolitical outlook is redolent of the old Nazi idea of dividing the world into pan-regions, each dominated by its own Great Power (Limes 1994). The central issue involves whether the public support for Zhirinovsky was a vote for such ideas or a protest against the economic reform process and its agents. Exit polls in areas where Zhirinovsky did well (generally regions with large military garrisons and decrepit industrial plants) showed no enthusiasm for foreign adventures (Steele 1994).

Unfortunately, the experience of totalitarian government left little basis for a renewed civil politics other than ethnic identity. Other social bases for political identity such as occupation or class are fatally associated with the class rhetoric of the old regime. Yet, the new ethnic nationalism (throughout the former Soviet Union and Eastern Europe) is fragmenting rather than imperialist. One of its strongest components is opposition to central economic control. The past state-formation – the Soviet Union for Russians, Yugoslavia for Serbs and Croats – has become the Other against which contemporary political identities are constructed (Kaldor 1993). This is not a Rip van Winkle-like revival of latent but primordial ethnic drives and hatreds (Eller and Coughlan 1993). It has immediate social roots in the collapse of centralized regimes that systematically depoliticized social identities and interests other than ethnic or 'national' ones. Indeed, there are strong decentralist pressures within Russia from ethnic minorities and regions that wish to enter into direct connection with foreign firms and governments. Even if the territorial integrity of the present Russian state is maintained, the vastness of the country may well lead, as a similar vastness has in the United States, to distinctive economic and political links to adjacent regions and countries (e.g. Karelia and Finland, the Far East and Japan).

The second factor is the degree to which the external world remains enticing rather than threatening. Liberalization of the Russian economy will require large infusions of foreign capital (Smith 1993). Given the global credit crunch in the early 1990s brought on by such phenomena as the US federal deficit, German unification and the collapse of the Japanese stock market, this may prove difficult. Membership of the European Community is a long way in the future. Several former satellite states in Eastern Europe – Hungary, Poland, and the Czech Republic – are much further down the path of meeting the economic and political criteria for EC membership than is Russia. A more immediate possibility is to form a trading bloc with other former Soviet states. More certainly, the relationship with adjacent states, particularly those formerly part of the Soviet Union, will be difficult given the mighty nuclear and conventional forces still under Russia's control. It has been tempting for the Ukraine to keep the nuclear weapons it inherited from the Soviet Union in order to have some leverage over Russia (Figure 6.1). But quite what specific global foreign policy goals, other than leverage over Western re-sources, the Russian nuclear forces now serve is unclear. They are much more useful in regional terms; bartering with the other former Soviet states and with Germany. They do guarantee a seat in the highest councils of inter-national organizations, but in light of fundamental economic weaknesses this is hardly indicative of a near-term hegemon-in-the-making. More than the Soviet Union collapsed in 1990. If the preceding analysis is correct, so too did the possibility of any new global hegemon from this region.

THE CHINESE CHALLENGE

Arguments for a 'Pacific Destiny' in the twenty-first century make much of the remarkable growth of the Chinese economy in the 1980s, at a time when the world economy as a whole, and the economies of poorer countries in particular, did not have high rates of growth. Pacific Destiny refers to a future world economy centred around the rim of the Pacific rather than the North Atlantic, with one of its principal geopolitical pivots in China. But China also figures as a future contender in conventional accounts of hegemonic succession. Its size in area and population and its long history of political integration are seen as important assets. As China develops economically it can be expected to become one of the truly Great Powers with possible hegemonic status.

With the largest population of any state in the world, 1.13 billion in 1990, China had a growth in GNP per capita between 1965 and 1990 that averaged 5.8 per cent per year (compared to a 2.9 per cent average for all low-income less-developed countries). In the 1980s its rates of growth in GDP, agri-culture, industry, manufacturing and services were among the world's highest. The rate of growth in manufacturing was particularly impressive: 14.4 per cent per annum between 1980 and 1990. Only Indonesia and South Korea came close to matching this figure among larger economies (World Bank 1992). As a result, China in 1991 was a little further along than South Korea had been in 1970 in terms of output per capita of some basic industrial products such as electricity, steel, cement, cotton fabrics and cotton yarn (Perkins 1992). With a fifth of the world's population China is now only a generation behind the older NICs of East Asia (Taiwan, South Korea) in

A

B

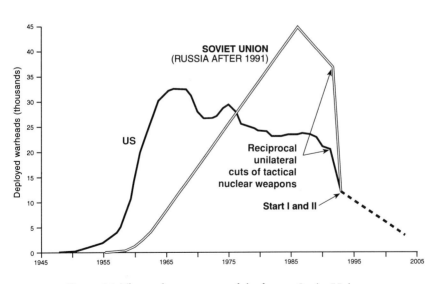

Figure 6.1 The nuclear weapons of the former Soviet Union
    A The location of weapons and disposal sites
    B The US and former Soviet nuclear arsenals, during and
       after the Cold War
Source: US Arms Control and Disarmament Agency 1992

141

conventional measures of economic growth (Table 6.1). Already China has the world's fourth largest territorial economy, after the US, Japan and Germany; approximately one-quarter the size of the US economy in 1992 at the prevailing market exchange rate. On the basis of a purchasing-power parity comparison (taking account of differences in the cost of living) China was the world's second biggest economy after the US in 1992 (*Economist* 1993b). In both relative (rate) and absolute terms, therefore, China is an economic colossus.

The lion's share of China's new growth has been concentrated in its coastal regions, especially in zones around Shanghai and Hong Kong that have been opened to foreign investment. Originally 'experimental', as the Communist leadership in Beijing tried to work out a new model for absorbing foreign capital, technology and management practices, the Special Economic Zones, such as Shenzhen on the border with Hong Kong, have become the nodal points for reforming the Chinese economy as a whole. This new model is not without political enemies, especially among those nostalgic for the autarkic China of the Cultural Revolution in which agriculture and national economic self-sufficiency took precedence over industrial development and interaction with the world economy. But since 1978, under the leadership of Deng Xiaoping, China has embarked on a total reorganization of its economy: decollectivizing agriculture, closing or privatizing parts of state-owned industry, dismantling central planning in favour of private entrepreneurship, and more fully integrating China into the capitalist world economy. Economic growth has replaced class struggle as the 'locomotive' of history.

The positive effects of reform have been strongest in coastal China and in the largest cities. In one account, the economic growth of the coastal areas represents the integration of Hong Kong and Taiwan with a set of continental hinterlands. Low labour and land costs have attracted investment from 'offshore' China. The surge of investment and growth has been led by ethnic Chinese business networks which stretch from South and East China to South-East Asia and around the 'Pacific Rim' to Canada and the United States.

Table 6.1 South Korea (1970) and China (1991), comparison of output per capita (US$ at official exchange rate)

Output	S. Korea (1970)	China (1991)
Electricity, kwh	291.6	579.9
Steel ingots, kg	15.3	60.9
Cement, kg	183.9	210.3
Nitrate fertilizer, kg	12.3	13.1
Cotton fabrics, m²	6.1	14.4
Cotton yarn, kg	2.9	3.9
Exports, US $	125	62
GNP, US $	1,100	322

Source: Perkins 1992

Economic imperatives have overcome political isolation. The link with Hong Kong is not too surprising given the imminent reintegration of Hong Kong with China in 1997. But Taiwan and China are still officially on a war footing, Taiwan being the refuge for the Nationalist government of China after the Communist Revolution of 1949. Although all economic relations are channelled through Hong Kong, in 1992 China became the biggest single destination for Taiwanese foreign direct investment, displacing Malaysia into second place. Pressure is mounting on the Taiwan government from Taiwanese business to allow direct trade and investment with the mainland. With the spillover of investment and trade to the mainland, China is following economically where Taiwan once led (Andrews 1992; Jones *et al.* 1993).

The euphoria of some commentators about the recent economic transformation of coastal China should be tempered by recognizing a number of barriers to the continuation of present trends without far-reaching institutional reform. One is the spatial polarization of economic development in China. The interior is quickly falling behind in the process of economic development. The very geographical pattern of economic development decried by the Communists in their ascent to power in the 1940s as the product of Western imperialism's subjugation of China is now being recreated under their aegis. The old 'treaty port' system of dual development – scattered development along the coast and on coalfields in the North with the mass of the population still engaged in agriculture – is in rapid reproduction. Knowledge of the disparities in incomes and opportunities is now much greater. The riots against 'unfair' and 'corrupt' tax collection in parts of interior China in 1993 are one not too surprising result. The central government may well have to intervene to prevent spatial polarization from generating even more extensive political conflict.

A second important barrier is the tension between local governments in the coastal areas and the central government. Local governments have acquired considerable autonomy in pushing credit expansion and investment. This has caused a severe overheating of the national economy as expanding credit chases a shrinking money supply controlled from the centre.

Third, and finally, for all the apparent success in creating a 'private' economy there is still a level of arbitrary government regulation without a guaranteed rule of law that restricts further foreign investment. Investors must worry constantly about maintaining the liquidity of their investments just in case the 'rules' change suddenly because a local bureaucrat or government minister shifts course.

At the root of all of these barriers is the lack of political change paralleling the Chinese economic change. Unlike the countries of Eastern Europe and the former Soviet Union which have undergone political change prior to economic reform, the Chinese government has created an increasingly capitalist economy within a still formally socialist state. The dominant view of many 'China watchers' is that the Chinese Communist Party is now largely irrelevant to economic life, although some institutional economists do maintain that 'development' is fostered by strong regimes of whatever type. What is clear, following the 1989 pro-democracy demonstrations, is that the Party's monopoly of state power is unacceptable to many Chinese citizens. The capitalist genie released in coastal China is probably too well-established now to push back into the lamp of Communist autarky. But quite how a strongly centralized state without a powerful Communist Party, and an externally-orientated economy with considerable decentralization, can be successfully combined remains to be seen (MacFarquhar 1992).

There has been a logical progression in Chinese official foreign policy in line with shifts in its relationships to both the world economy and to other states. In the 1960s the Chinese government espoused revolutionary change within states as the means of bringing about a global political transformation. Lin Biao's countryside versus city analogy, conveyed most clearly in his paean to people's war in September 1965, embodied such thinking in its purest form. By the mid-1970s, as China expanded its diplomatic dealings, the emphasis was increasingly placed on the need to reform international economic relations rather than change the internal political character of states. In the early 1980s, as the Chinese government committed itself to long-term relationships with the non-Communist industrialized world, Soviet support for wars of national liberation was repeatedly castigated. China has thus moved from a globalist to a statist conception of development and security.

In aspiration as well as in fact China also remains a regional rather than world power. The impact of Chinese economic and military capacities are felt along China's periphery, not elsewhere. The strongest economic ties are the regional ones. The main foreign policy issues are relations with Vietnam, Taiwan and the two Korean states. Beyond these issues lurk problems of territorial sovereignty and claims to ocean-based resources. A Chinese map of China serves as a reminder of the disputed and unresolved claims between China and its neighbours. In this context, the depiction of China as a reactive, defensive state is credible only in relation to the United States and Russia. In Asia China's military power and repeated willingness to use it have begun to fuel a significant arms build-up among adjacent states. China, Taiwan and the two Koreas are all developing chemical weapons and ballistic missiles. This investment in military competition reflects in part the failure of political change to follow economic development, but it is also the result of economic worries. To finance further economic growth these countries are all concerned about exploiting their offshore resources (oil, fisheries, etc.). However, the boundaries of the offshore regions are overlapping and contested, so there is a growing risk of territorial conflict. Ironically, increased integration into the world economy has encouraged an increase in East Asian territorial disputes as states compete for regional resources (Klare 1993).

China's economic growth, geographical scope and emergence as a leading actor in East Asian politics seem destined to make China an arena of importance in world politics. However, there are a number of potential drags on the movement to Great Power status. One is the degree to which China can accommodate the economic practices of the Asian NICs without putting at risk the authority of its central government. A second is that China and its population will remain largely agrarian until well into the twenty-first century. The awesome tasks of feeding, clothing and housing China's billion plus people will set limits on the effects of external economic ties. China is not simply its coastal provinces. This raises a third question: much of the recent economic growth of China relies on foreign investment and exporting manufactured goods. These also set limits on the pursuit of political primacy, even in East Asia. It is political 'enemies' such as Taiwan that provide much of the capital for Chinese economic development. The United States, Japan and other industrial countries provide many of the markets for Chinese exports. This export-oriented type of economic development does not lend itself to a conventional political–military primacy. The late Chairman Mao would have recognized this as an unresolvable contradiction of the Faustian bargain that China has struck with the capitalist world economy.

THE JAPANESE ENIGMA

By most indicators Japan has the second largest economy in the world and, until their recent decline or stagnation, the highest rates of growth in productivity and output. As a consequence it is seen as the most likely of the present Great Powers to challenge as a hegemonic successor to the US. Certainly this is the view in both the United States and Japan, where large academic–publishing–entertainment enterprises have sprung up advertising *The Coming War with Japan, Rising Sun*, or *The Japan that Can Say No*. In this context the significance of Japan's 'rise to economic power' has remained difficult to assess. Scenarios range from Vogel's fully-fledged 'Pax Nipponica' through Bergsten's US/Japan condominium or 'joint hegemony', to Rosecrance's view of Japan as the prototypical 'trading state' inaugurating a new world economic order (for a survey see Inoguchi 1988–9).

This confused state of affairs can be explained partly by differences between commentators over how they interpret the nature of economic power, i.e. can it be an end in itself or is it always a means to other political–military ends? But it is more the result of how 'Japan' itself is understood. Japan was the first culturally non-European Great Power in the modern state system. Its success first as a territorial empire in Asia, and more recently as a global economic actor, has been difficult for Europeans (and Americans) to explain and accept. The trajectory followed since the devastating military defeat of its empire in 1945 defies both of the most popular explanations for economic success: (1) land area/resource endowments and (2) comparative advantage and specialization. Despite its poor resources Japan has established competitive advantages in a wide range of manufacturing and service activities that serve global as well as Japanese markets.

Two 'special' explanations have been favoured in accounting for Japan's success, each with a different prognosis for future development. On the one hand, the Japanese economy has been seen as working in compliance with one or another economic theory that has given it advantages over countries less faithful to a particular economic wisdom. On the other hand, Japan is viewed as institutionally deviant, organized in such a way as to exploit a world economy that works by different rules. Compliance is suggested by the OECD's attestation to Japan's Keynesian demand management, by Milton Friedman's attestation to its control over its money supply, and by Arthur Laffer's attestation to its low corporate tax/high industrial investment, to identify only a few examples. In each case Japan is regarded as an economic entity without any attention to its potential as a political actor.

In contrast, two different kinds of deviation are identified for either condemnation or admiration. One is the purported 'plan rationality' as opposed to 'market rationality' upon which Japanese institutions are based (e.g. Johnson 1982, 1987). MITI, the Ministry of International Trade and Industry, is frequently identified as the central pillar for a 'Japan Inc.', systematically identifying leading industrial sectors and planning their global success. The Ministry of Finance is sometimes also given an influential role. The dominant political party since 1955, the Liberal Democratic Party (LDP), is viewed as the main link between big business and other entrenched economic interests (such as rice farmers) and the bureaucracy. A second instance of 'deviation' refers to an absence of strong leadership in a system of informal alliances between politics, business and bureaucracy (e.g. van Wolferen 1990). From this point of view:

> The political give-and-take among the System's components interferes with the nation's need to deal with the rest of the world. To take only one example, many if not most Japanese politicians have an un-acknowledged stake in what is known euphemistically as *boeki masatsu*, or trade friction. They want to keep it from doing too much damage, and they ask the bureaucrats to take limited 'cosmetic' action. But if the issue disappeared so would a lucrative source of income, since the sectors of the business community that most fear foreign competition in the home market are also the most generous with 'political donations'.
>
> (van Wolferen 1990, 48–9)

In both cases Japan is 'unprecedented and dangerous' in its possible mutation into an economic–military superpower.

The difficulties of these approaches are apparent. Not all of the modes of compliance are compatible with one another, yet there obviously is compliance to certain norms of economic conduct if the judgements of economists have even a modicum of validity. Moreover, it is not so much institutional deviation that accounts for the Japanese 'difference', as adherence to a distinctive type of economic action, adapting pragmatically to problems as they arise but within a framework provided by elements of several economic theories. Schmiegelow and Schmiegelow (1990, 557) term this 'strategic pragmatism'. Detailed analysis of Japan's external relations has suggested persuasive evidence of both a high degree of ability to adapt to the world economy and a reluctance to expand non-economic power capabilities (Calder 1988; Schmiegelow and Schmiegelow 1989, 1990). Neither of these attributes is compatible with either the analysis or the fears of the deviationists, for whom Japan is always proactive rather than reactive. But it is the model of Japan as a reactive state that seems to better fit the evidence.

Consider first of all adaptation to the world economy. Since the mid-1970s, as its annual rates of productivity growth have declined, the Japanese economy has nevertheless become the world's second largest economy and its leading creditor. This is because compared to other industrialized countries Japan has maintained a more competitive economy across a wider range of industries. It has done so as a result of adapting successfully to outside pressures rather than as a result of intrinsic advantages that flow from its political–economic peculiarity (Leyshon 1994). Japan is proving better able to cope with the globalized world economy than many other countries, more than posing a threat to it. Its historic vulnerability has prepared the country for adaptability. Several pieces of evidence support this interpretation.

First, from 1950 until 1970 Japanese trade policy can be described as having involved a combination of export promotion with import substitution and high value-added industrialization (Tyson and Zysman 1989). Since the early 1970s, however, significant trade liberalization has occurred. As part of the negotiations on the reversion of Okinawa to Japan, then still occupied by the United States, Japan agreed to begin the removal of tariff and investment restrictions. Further efforts at removing barriers to trade have followed. The most thorough surveys find little evidence for much difference in the levels of protection between the US and Japan (e.g. Bergsten and Cline 1987). But because some of the Japanese barriers, particularly those involving licensing approval, are less 'transparent' than American ones, there is a persisting belief in the US that Japan does not 'play fair' on trade. Certainly moves towards free trade have not been without domestic political consequences in Japan, as some parts of the bureaucracy, particularly the Ministry of Agriculture and Fisheries, have consistently opposed the opening of Japanese markets (Grant 1993). Capital liberalization has also proved more difficult than trade liberalization (see, for example, Figure 5.7). Only when US multinationals combine with Japanese business has there been much success in this area. Pressure from the US government and international organizations has not been as successful as with trade policy (Encarnation and Mason 1990).

Second, the adjustment of the Japanese economy to the 'oil shocks' of the 1970s was much smoother than that of many other industrialized countries. But Japan avoided a mercantilist response; there was no attempt at turning away from a world economy which supplied the bulk of Japan's energy needs

(Ikenberry 1986). Adaptation was mediated by domestic actors such as Nikkeiren, the business association, and private-sector union leaders. It was not the state alone that resolved the dilemmas facing the Japanese economy (Kume 1988). It was, above all, a new domestic coalition which gave private-sector unions a key role that determined the successful Japanese resolution of the oil crisis. Indeed, without this political adaptation the Japanese response, particularly in 1973, would have been much more likely to fail (Volcker and Gyohten 1992). The world economy, therefore, again had important effects on the Japanese political economy. So, though the 'strength' of the state in Japan contributes to the adjustment capacity of the economy, 'the international system has a continuing impact on Japanese domestic politics, and from time to time it forces changes in the nature of politics' (Kume 1988, 685).

Third, Japan's external dependence for most industrial raw materials and many foodstuffs has led to an enhanced concern with 'economic security'. This is something many American commentators, coming from a country with a history of resource independence, do not seem to understand (see, e.g. Vernon 1983). The perception of vulnerability leads many Japanese to accept higher domestic prices in return for long-term contracts with external suppliers and domestic production (Samuels 1989). Rather than a sign of strength and autonomy in relation to the world economy, therefore, the 'notorious' (in the eyes of the American deviationists) privileging of production over consumption in Japan is much more a sign of both economic weakness and political responsiveness to external dependence.

Fourth, the Japanese government has been responsive to external pressures to the extent that normal macroeconomic adjustment measures, particularly exchange-rate adjustments, do work as well, if not better, in the US–Japan relationship as in the US–Europe relationship (Cline 1993). For example, in 1986–7 the Japanese government responded to American prodding to stimulate domestic consumption with policies that had the effect of raising urban land prices in Japan to astronomical levels. This was the price the Japanese economy paid (doubled when the prices collapsed in 1992) for helping to rebalance the rapidly unbalancing US–Japan trade relationship. But this was a reactive action, not a conspiracy by the Japanese business élite in cahoots with the Ministry of Finance and factions of the governing LDP to build a 'wondrous money machine' designed to buy up such American business icons as Columbia Pictures and the Rockefeller Center in New York (van Wolferen 1989).

Weakness can be turned to strength, therefore, under certain conditions. But a number of constraints will limit the possibility of ever turning this reactive capability into a singular 'hegemony'. One is that Japan is geographically isolated from its principal industrial markets. The Japanese reincorporation into the world economy after the Second World War involved a reorientation from adjacent Asian countries to the larger industrial capitalist world beyond Asia. Although Japanese trade and investment in Asia have increased substantially since the 1970s, the orientation of the Japanese economy is over-whelmingly towards Europe and North America (Grant 1993, 285). A second constraint is Japan's heavy external dependence on sources of raw materials. Japan is the number one importer of basic raw materials (petroleum, iron ore, copper ore, etc.) among all the industrial countries in the OECD (GATT 1990). It is also the world's largest importer of agricultural goods. Unless it can obtain raw materials and foodstuffs through territorial expansion, the course of action that led to the disasters of the Second World War, Japan must remain committed to an open world economy to buy its raw materials and to sell its manufactured goods. Third, contrary to much conventional opinion, Japanese strategies for new product innovation have not been uniformly successful. Only in some industrial sectors, particularly engineering and electronic components, has Japanese technological innovation been superior to that elsewhere. One expert concludes 'it is not clear that Japan offers the most advanced and efficient institutions adapted to the present innovation frontier, as Japan enthusiasts have often maintained' (Kitschelt 1991, 489). Fourth, and finally, Japan faces a demographic 'trap' much more severe than that of other industrial countries. Not only is its population over age 65 growing at an extremely rapid pace (Martin 1989), but Japan remains extremely 'unfriendly' to immigration as a solution to the labour shortages it is likely to face in the near future. In response to this dilemma Japanese business will probably have to increase the level of investment outside of Japan so as to obtain access to labour (Jones 1988). This will further increase the inter-nationalization of the Japanese economy in the future.

A reluctance to turn its new economic strength into military capacity is also a feature of contemporary Japan. From one point of view this is because the alliance with the US still has considerable payoff for Japan in both lower military expenditures and a low profile in international politics that allows Japanese business access to all kinds of countries including those with 'pariah' governments (e.g. South Africa, North Korea). This could change if the US withdrew from its offensive military position in North-East Asia. If this happens some alarmists foresee a brutal struggle between the USA and Japan as both replay the Second World War all over again (the most alarmist is Friedman and LeBard 1991). What this scenario and other more modest ones predicting increased defensive military capabilities ignore, apart from the nature of the economic ties of Japan to the world economy, are (1) a Japanese security policy formulated within institutions that are biased against military security objectives and (2) deeply entrenched popular attitudes opposed to rearmament and an active military presence beyond the boundaries of Japan.

The principal transformations in the global geopolitical order since the mid-1980s – including the rise of Japan as an economic creditor and the collapse of the Soviet Union – have not elicited any dramatic change in Japanese security policy. Domestic institutions have persisted in the course established during the American occupation of Japan (1945–52) with the limited exceptions of increasing participation in foreign aid programmes (usually multilateral) and, most recently, contributing forces to UN peacekeeping duties in Cambodia (Weinstein 1971; Harries and Harries 1987). This course has been to define security in broader economic and political terms, rather than exclusively, as in the US, in military terms. In particular, the main mechanisms of policy-making do not encourage the definition and implementation of military objectives by either the Japan Defence Agency or the Prime Minister. Each is subject to severe controls and legislative oversight which seriously limits their capacity to develop military goals and plans (Katzenstein and Okawara 1993). Neither is there any electoral return to advocating a more unilateral military policy, let alone a military build-up. As Cowhey (1993, 321) shows, 'no political leader benefits significantly (nor does the LDP as a whole) from dwelling on issues beyond patronage politics or commercial diplomacy to fuel export expansion.' Though the slow disintegration of the LDP may change this with the emergence of politicians such as Ichiro Ozawa openly committed to greater Japanese involvement in foreign–military affairs, there is as yet no large constituency for a remilitarized Japan.

The growth in Japan's military spending in the 1980s was substantial (an annual average of 6.5 per cent growth in the defence budget) but it was conducted in close collaboration with the US military and is designed to complement American forces rather than give Japan an independent opera-tional capability. Most of Japan's industrial leaders remain opposed to expanding arms production. Most importantly, the Japanese public appears largely unimpressed by arguments about 'meeting aggression or defending the principle of national sovereignty' (Berger 1993, 129). Apart from the continuing dispute with Russia over the (formerly Japanese) Kurile Islands to the north-east of Hokkaido, Japan is itself involved in no significant territorial disputes.

The roots of popular opposition to rearmament lie in Japan's total defeat in the Second World War and the key role an uncontrolled military played in bringing about that outcome. Unlike in Germany, where a variety of social forces were involved in the rise of Nazism and the country's subsequent involvement in the Second World War, in Japan the war is widely seen as having been precipitated by a militarist government without much popular support. The sense is of a Japanese population victimized by its own military as much as by foreign states (Berger 1993, 136). One consequence of this has been a lack of national self-recrimination over the Japanese role in starting the war and with respect to the charges of war crimes made by Chinese and other Asian governments. This continues to fuel suspicion of Japan through-out Asia and Japan has only recently showed signs of coming to terms with it. Another, however, is a persisting negative view of the military shared all across the political spectrum (Berger 1993, 137). Though groups differ on how they propose to prevent the military from again becoming a threat, they agree on the need to restrict the ability of the military to plan or organize any kind of offensive capability. Japan's economic success since 1945 is widely seen in Japan as having depended on anti-militarism. As the American saying goes: 'If it ain't broke don't fix it.' For the foreseeable future this is how most people in Japan will see it too.

GERMANY, AGAIN?

Germany was the state that persistently challenged the geopolitical status quo during the long period of inter-imperial rivalry (1875–1945). This challenge, expressed first in the disciplined attempt at building an industrial state and a colonial empire, and, later, in the establishment of the Third Reich, was defined by the French, British and Americans as 'the German problem'. The dilemma was: how could other states long accustomed to greater power accept as an equal a recently unified state with a large population, substantial industrial strength and high standards of technical education, without sacrificing their own status? Dealing with this was the chief feature of world geopolitics from 1875 to 1945, as described in Chapters 2 and 3.

As a result of Hitler's decision in 1941 to go to war with both the Soviet Union and the United States, the German problem was resolved in a fashion after 1945 through the *de facto* partition of Germany. The Cold War was institutionalized in large part through the division which ran between the Western and Soviet zones of occupation. The great fault line of American–Soviet enmity ran through Berlin, the historic capital of a unified Germany. The two states which evolved from the military occupation by the erstwhile allies of the Second World War appeared by, say, 1980 to be a permanent feature of the world political map.

This stability might have endured for generations but for the unravelling of the Soviet sphere of influence in Eastern Europe from 1985 to 1989. Soviet President Gorbachev was the sorcerer's apprentice. At a stroke, he resurrected the possibility of a new German problem when he consented to the opening of the Berlin Wall in 1989. The Wall, in place since 1961, had become the principal landscape symbol of the territorial division of Germany. Its destruction has become the principal sign of a new Germany, much more likely than its disunited parts to aspire to the hegemonic status so often denied to it in the past. This fear motivated many Germans as well as other Europeans to anguish over the rapid pace of unification, pushed by East Germans and the nature of the Soviet–West German Accord of June 1989 (McAdams 1990), and to look to the deepening of political–economic ties within the European Community as a way of restraining the German Gulliver.

The spectre of the past is also readily conjured up in one of the most immediate consequences of reunification: the growth of neo-Nazi extremism and violent attacks by 'skinhead' gangs on immigrants and asylum-seekers. This coincided with a severe economic downturn in Germany as a whole, high unemployment in the former East Germany, and a serious crisis within the European Community over the prospects for further integration (see next section). Attempts by some well-known historians and politicians in West Germany in the years prior to reunification both to equate the crimes of the Nazi regime with those committed by other Great Powers, and rehabilitate Nazism as a struggle against communism, did little to inspire confidence that history produced lessons rather than repetition (Evans 1989). Indeed, inadvertently or not the Reagan administration reinforced the acceptability of the view that Nazism was the expression of Hitler's personal will more than a phenomenon in which most Germans had either actively participated or passively accepted when, in 1985, President Reagan chose to attend a ceremony at a cemetery in Bitburg, West Germany in which members of the fanatically Nazi Waffen-SS were buried (Mommsen 1987; Hartman 1986). The President's message at the ceremony was that the SS soldiers were victims of Hitler as much as the groups who are usually identified as their victims – Jews, Gypsies, Slavs, etc. This revisionism fits neatly with the goals of those who want to believe that Nazism died with Hitler. It also worries those who see the Germans failing to come to terms with a past that the aftermath of reunification has served only to highlight. If the federal, liberal and restrained West Germany could be cast in the role of the Weimar republic, then reunification and its aftermath can be read as a replay of the rise of Hitler and the Nazis. Only the new Hitler is missing, so far.

Is this the direction in which Germany is now headed? The Cold War centred upon the division of Germany. The war plans of both sides saw Germany as the first battlefield in World War III. This was the ultimate discipline that the victors of the Second World War exercised over the two German states. The reunification and demilitarization of Germany offer the possibility of an independent German presence in world politics for the first time in fifty years. The NATO alliance that kept West Germany as an integral part of the Western–American military enterprise has lost its reason for existence with the collapse of the Soviet sphere of influence. This is indicated above all in its inability to organize a collective response to the disintegration of the former Yugoslavia in 1990–1; Germany followed an independent course in recognizing the independence of Slovenia, Croatia and Bosnia-Herzegovina but without the political–military capability to back up its policy (Treverton 1991/2). This helped set the conditions for the dreadful inter-ethnic wars of 1991–4 in Croatia and Bosnia to which NATO had offered no effective response by 1994 (see also Denitch 1993). Those hoping to rescue NATO as a continuing representative of 'the West' need to account for this dramatic failure (e.g. Glaser 1993). This 'West' is an increasingly fictive entity at least as far as European collective security is concerned.

Yet, there are good reasons for suspecting that the breakdown of the Western (Atlantic) Alliance will not initiate the Second Coming of the Third Reich. First, the institutions of modern West Germany have taken deep root (Paterson and Southern 1991). The analogy to Weimar is misleading. The political system is decentralized in many important respects and loyalties have been built to localities and regions as much as to the state as a whole. The process of reunification has made this clear. It is in the eastern *Länder* (of the former Communist DDR) that extremist right-wing politics has had the most wide-ranging support. There it may take several generations to establish the political imagination that years of subjugation to propaganda and secret police surveillance effectively destroyed. Those who even a few years ago could equate the two systems as similarly despotic, such as the former leaders of the West German Greens, have long since lost their political credibility.

Second, the costs of reunification have made the German government turn inward. The internal challenges will absorb the government's energies for several years. The first of these is responding to the collapse of much of East Germany's uncompetitive economy. The second is maintaining fiscal stability in the face of the huge capital needs involved in rebuilding the East, cleaning up the various environmental disasters there, and meeting the sudden surge in pension and welfare demands (Giersch 1992; Welfens 1992). The fateful decision to unite the currencies of the two Germanies at parity has had the unfortunate effect of encouraging the central bank, the Bundesbank, famously protective of the value of the German Mark and West Germany's low rates of inflation (Goodman 1992), to maintain high interest rates to fund reunification. This had disastrous effects on the European Exchange Rate Mechanism (a vital prop of any future European Monetary System) when the Bundesbank refused to lower interest rates in order to prevent speculative attacks on the British pound and Italian lira (1992) and French franc (1993). Perhaps nothing is as symbolic of the 'new Germany' as the priority given by the Bundesbank to re-establishing fiscal stability, even if this is at the expense of German influence or goodwill in Europe as a whole. This perhaps has its historic roots in the experience of the 'Great Inflation' after the First World War which seems to have become an integral part of German fiscal policy (Feldman 1994).

Even before reunification, however, there were signs of stagnation in West Germany's 'model' economy. Many of the *Mittelstand* of small and medium-sized businesses, which form the core of German industry, had lost their competitive edge. In the 1980s worker productivity grew faster in the US, Britain, Japan and France than it did in West Germany. By 1991 German labour costs per unit of output were 36 per cent higher than in Britain and Japan and 45 per cent higher than in France. Attractive and reliable German products could no longer be relied upon to offset such a handicap. Other countries, including the Asian NICs, could now offer quality products more cheaply. High wage settlements, the costs of an extensive social benefit system, and the impact of a highly-valued currency on the growth of exports, conspired to limit the capacity of the 'engine' of European economic growth.

Third, Germany's central position in the European Community is important to the future development of either a more integrated Europe or an expanded Community, in whichever direction it goes. As the largest national economy within the EC, Germany has extensive trade and investment links with all of the other members. It is also the most involved with trade and investment in Eastern Europe. Politically the most significant tie is with France. Since the late 1970s, and despite the monetary discord noted previously, trade and investment have blossomed between these two 'traditional enemies'. The two countries were each other's biggest trading partners in 1992, France taking 13 per cent of German exports, and sending back 18 per cent of its own. Even more striking has been the boom in cross-border investment. It is most visible in the old East Germany where French firms have made or promised bigger investments than those from any other foreign country. German firms are second only to American firms as investors in France. There are also numerous Franco-German joint ventures in power plant construction, banking and aerospace (*Economist* 1993a). The German economy is firmly anchored in a European economic space.

Fourth, and finally, the analogy with the Nazi past misses two fundamental political changes since that time. One is the diminution with passing time of the territorial claims that served to incite so much passion in the past. Most Germans (if Austrians and Swiss Germans are excluded) are now within the boundaries of one state. This was not the case in the past. Though there are still those on the far right, often refugees from the Sudetenland, former East Prussia, and other 'lost territories' in the East, who make claims to a larger German *Reich*, there are few ethnic Germans left in those regions following the flight and clearance of the years immediately after the Second World War. More importantly, rather than the desire for *Lebensraum* (living space) in Eastern Europe, the contemporary German demand is for erecting barriers to immigration from the ex-Comecon countries. The experience of incorporating the former East Germany has further tarnished the attraction of territorial aggrandizement within the former Soviet sphere of influence.

Another change involves the acute consciousness of the past manifested by many Germans despite some recent attempts at historical revisionism. Germans are constantly worrying about the stability of their institutions and the 'last time' they collapsed. The sense of being *nazistisch belastet* – burdened by National Socialism – can appear almost pathological. Wanting to put this past behind them and make Germany an *ordinary* country again is what inspires much historical revisionism. But the popular consciousness of the past remains a barrier both to the acceptance of a more nationalist history and a return to an activist foreign policy. One essayist has put this in a nutshell, writing of West Germany before reunification:

> No one is more suspicious of the Germans than the Germans themselves. You cannot pass a week in the Federal Republic without seeing at least one article asking, 'How Stable Is Our Democracy?' or, 'Is Bonn Weimar?' or, 'How Open Is the German Question?' West Germany is like one of its own model businessmen: a hearty, suntanned forty-year-

old, hair neatly parted, smartly dressed (Harris tweed jacket, gray flannels), with nice manners and a stock of sensible conversation – but forever dashing into the Apotheke to check his blood pressure, or glancing at his reflection in the windows, to see if he hasn't developed a nervous tic. To nations with happier pasts, this relentless self-examination may seem excessive or even faintly comical; but we would really start worrying if Germans stopped worrying.

(Ash 1990, 72–3)

TRADING BLOCS

One of the leading political–economic innovations of the Cold War period was the establishment of the European Economic Community (now European Union, EC in the text) in 1957 and its expansion over the next thirty years to include twelve European countries in total. Inspired by the ideal of overcoming historic political antagonisms through economic integration, the EC was also encouraged by the United States as yet another means of combating Soviet influence in Europe. The apparent success of the EC in stimulating economic growth in Europe has recently inspired the proposal of an enlarged North American Free Trade Area (NAFTA) covering Mexico as well as the US and Canada, which already had an agreement dating from 1988. In East Asia the flood of Japanese investment, and the increased use of the Japanese yen as a 'regional currency', have led to characterizations of the region as an incipient 'trading bloc'. Elsewhere, in Latin America, Africa, South-East Asia and Australasia, regional trading agreements have grown spectacularly since the late-1970s.

Though each of the three principal 'entities' (EC, NAFTA and East Asia) is of a quite different type and at a different stage of organization, some commentators have begun to see a world economy forming around a geographical 'triad' which, they claim, will provide the basic structure for an emerging geopolitical order (e.g. Wallerstein 1984; Gilpin 1987; Padoa-Schioppa 1993). Other arrangements can also be foreseen, such as a 'new bipolarity' with a greater Europe on one side and a 'Pacific World' (the US, Japan and the Pacific Rim) on the other. Either way, a world of regional trading blocs is the main alternative envisaged to the pursuit of primacy by existing Great Powers. In this construction the main actors are world-regional blocs (and dominant states within them) attempting to create territorial 'monopolies' through geographical accretion and market integration.

What are the characteristics of these 'blocs'? Do they form the basis for the territorial 'trading fortress' role they are being given? The 'tripolar' view of the world, with the United States, Japan and the EC as the three poles, has attracted the imagination of many commentators (see, for example, Aho and Ostry 1990; Lorenz 1991; O'Loughlin and van der Wusten 1990; Thurow 1992). The three poles are the biggest traders in the world economy and each

155

is the centre of trading for other countries based on geographical proximity, trade preferences and investment flows. In its most restricted form the reasoning is that the blocs as they stand have incentives to restrict trade with other parties. In a more elaborate formulation the blocs become akin to the 'pan-regions' of the Nazi geopoliticians dividing up the world economy into three world empires, with three Norths and three Souths: the US with the Americas, Japan with Asia and the Western Pacific, and the EC (Germany in the original) with Eastern Europe, the Middle East and Africa (O'Loughlin and van der Wusten 1990; also see Chapter 3).

The incentives to regionalize trade that are identified run the gamut from the narrowly economic to the political and cultural. The first is that protected industries can obtain larger markets and achieve economies of scale, yet still shield themselves from global competition. A second is that increased intra-regional trade serves to encourage other types of economic collaboration that produce regional gains (particularly for those businesses operating across many borders within the region), such as currency union, joint legal and regulatory environments, common quality standards, etc. A third is the frustration with the course of global-multilateral attempts at liberalizing trade, such as the GATT. It is usually easier to reach agreement when fewer parties are involved in negotiations. A fourth incentive, especially important in the case of the EC, is the hope that economic integration will strengthen political ties and reduce the likelihood of conflict between member states. Sometimes associated with this is the idea of building a common regional political identity that will finally subdue the competitive nationalisms that afflicted Europe earlier in this century. Finally, regionalization takes on its own dynamic once under way. Countries seek membership in preferential trading arrangements not so much because they expect significant gains but more because they wish to avoid the losses from trade diversion to which they would be subject as outsiders. In other words, there is a 'bandwagon' effect associated with regional trading blocs; once established they attract new members because of the potential costs of exclusion (Fieleke 1992).

In terms of the geography of trade and foreign direct investment flows, it is clear that the world economy is organized around a 'triad' of economies in the shape of the EC, the US and Japan. At first sight the EC is the most impressive of the three, with respect to population and GNP at current exchange rates (Table 6.2). But the US is ahead on GNP at purchasing-power parity and with respect to stock-market capitalization, the number of big corporations and energy consumption. If the EC ever becomes a 'single country' it would outvote the US in the IMF and the World Bank. In sporting and intellectual contests (the summer Olympic Games of 1988 and recent Nobel prizes) the EC and the US split the honours; Japan lags behind.

The 'triad' is central to world trade (Figure 6.2). By the late 1970s a trading system defined by the three 'cores' and their associated economies had come into existence. Each is hardly a separate world unto itself. Rather, associated

Table 6.2 Which 'bloc' is number one?

1990 figures	United States	EC	Japan
Population, m	251	328	124
GNP: current exchange rates, $bn	5,391	6,010	2,942
PPP exchange rates, $bn	5,391	4,664	2,130
Exports*, $bn	393	526	287
Energy consumption, mtoe**	1,974	1,191	436
Foreign-exchange reserves by currency, % of total	51	37†	9
International bond issues by currency, % of total	38	41	13
Stockmarket capitalization August 1991, $bn	3,350	2,040	2,841
Number of companies in Fortune 500	164	129	111
Government debt, $bn	1,796	2,022	874
IMF voting power, %	19.6	28.9	6.1
World Bank voting power, %	15.1	29.7	8.7
Olympic medals summer 1988, % of total	13	29	2

Nobel prizes, % of total
    sciences (1957–90)
    economics (1969–90)
	53
	60
		30
		23
			2
			0

* excluding intra-EC trade
† including ecu balances
** million tonnes of oil equivalent
Source: OECD; BP; IMF Morgan Stanley Capital International; *Fortune*; Salomon Brothers; World Bank; *World Almanac*; Nobel Foundation

economies are related to the main inter-triad circuits of the world economy through their nearest core. The triad also accounts for the lion's share of world foreign direct investment. The three cores accounted for 81 per cent of total world outward FDI in 1988 compared to 47 per cent of world exports. They are each increasingly the target of the others' FDI; investment between the three rose from 30 per cent of world FDI in 1980 to 39 per cent in 1988. Investment between the US and the EC accounts for four-fifths of this. So they are hardly isolated from one another. But FDI to other countries is largely dominated by the triad member of a particular region. For example, American firms contribute 61 per cent of all FDI in Mexico, Japanese firms 52 per cent in South Korea. Being closed out of these regional links increasingly means being closed out of the growth areas in the world economy (*Economist* 1991).

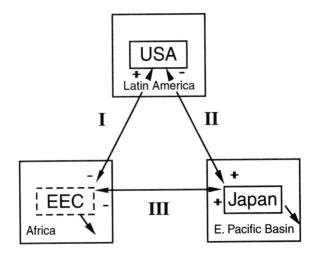

Figure 6.2 The circuits of world trade in the late 1970s

Of the three, however, the EC (European Union) is the only *true* trading bloc, having begun in 1957 with six members (France, West Germany, Belgium, Netherlands, Luxembourg and Italy) and expanded from 1972 onwards to a total of twelve member countries in 1993 (add in Britain, Ireland, Denmark, Greece, Spain and Portugal). From the outset the Community has exercised a pull on adjacent countries still outside it. The history of EFTA is instructive. The European Free Trade Association was founded in 1960 as a different way to European economic integration – freer trade without surrender of sovereignty to central institutions. Some of its original members have defected to the EC, and the seven that remain have negotiated an associate membership which should lead to full membership. Many countries in Eastern Europe are desperate to join at the earliest possible opportunity.

From the 1970s until 1986 the Community failed to evolve politically. But a new grand design instituted in 1986 aimed the EC towards a single market in 1992, and the Single European Act (1987) destroyed the principle of a national veto that had previously paralysed the Community's rule-making. With new momentum the EC headed towards the ideal of a Europe with one currency and one central bank. A deepened European market was seen by France and West Germany, in particular, as the way out of their poor economic performances in the late 1970s and early 1980s. European multi-nationals were also active in pushing in this direction (Garrett 1992; Cowles 1993). The Maastricht Treaty of 1991, inspired by a French desire to tie Germany irrevocably to a united Europe, pushed towards political union (Sandholtz 1993). Though market integration through monetary union re-mained vital to the conception of European integration, the Maastricht Treaty included provisions for the imposition of common standards governing social

158

and economic welfare. Though the British insisted on opting out of this requirement, it was included to give a sense that more was at work in producing the treaty than purely economic calculation. There certainly was no pressure from organized interests within the member states to include the so-called social protocol (Lange 1993).

The glitzy reputation of the EC among some commentators (especially Americans such as Thurow 1992), should not be allowed to obscure the real problems and dilemmas that it faces. First, the economies of the EC's member states are still a long way from the state of 'convergence' (on interest rates, budget deficits and the like) defined in and demanded by the Treaty of Maastricht. Most devastating of all, the ERM has all but collapsed as a way of aligning currencies. There seems little prospect of achieving European monetary integration across all EC members before the turn of the century, if at all.

Second, the general trend in the economic performance of the EC is downwards. Even incorporating the cyclical trend of declining growth rates throughout the industrialized world since the early 1970s, the net movement in the EC's real GDP (percentage change from previous year) has been downhill since Britain joined in 1973 (Figure 6.3). This trend cannot be written off as a recessionary effect. The expanded EC does not deliver the high rates of growth that characterized the original EC.

Third, the EC is increasingly less a single enterprise than a set of overlapping activity fields. While this maintains diversity it also reduces the possibilities for concerted action across a range of activities. For example, the ERM now includes less than half the EC membership; Britain and Denmark have opted out of several of the Maastricht accords; even the Schengen passport-free zone has only eight members. The net trend is towards a Europe à la carte.

This is symptomatic of a fourth problem: the appearance of an EC 'core' economy with a 'second' EC increasingly 'peripheral' to it. In monetary terms there is the real possibility of Germany, France, Denmark, Belgium, the Netherlands and Luxembourg (plus the EFTA members Austria, Sweden and Switzerland) forming some sort of union. However, it is unlikely that the other members will be able to participate (Bayoumi and Eichengreen 1992). Indeed, a case can be made for the view that such national economies as Poland, Hungary and Czechoslovakia (prior to its division) are better candidates for integration into the single market than Italy or Greece (*Economist* 1992). One effect of this would be to strengthen the German hand within the EC relative to France, given the high degree of incorporation of these countries into the German economy.

Fifth, two competing models of financial organization are implicit in the Single Market programme and the Maastricht Treaty. On the one hand is a liberalization model associated with British practice, in which there is light regulation and competition between financial institutions. On the other hand is a consolidation model associated with Germany, in which financial

159

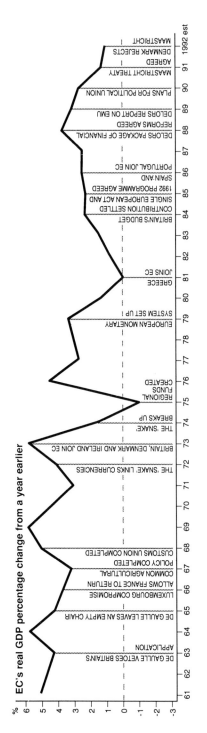

Figure 6.3 The downward trend of EC GDP, 1961–92
Source: OECD National Reports 1993

institutions are closely involved in managing industries and typically take a longer view of their investments. The former lends itself easily to cross-border capital flows but is likely to be resisted in Germany because of the threat it poses to the German model of economic development. If it succeeds, it will make monetary regulation within the EC as arduous as it has been in Britain (Leyshon and Thrift 1992).

Sixth, and finally, the collapse of communism may well also signal the collapse of further moves towards European union. The most striking feature of Europe in the early 1990s is the feebleness of European consciousness. Most of the member countries of the EC are caught up in their own distinctive affairs. Save for hostility to immigrants there is little that unites Europeans politically. The many different responses of the EC to the Bosnian crisis of 1991– are only the most obvious example of the absence of any unified European consciousness and political will (Hodgson 1993).

As a potential Great Power, therefore, the EC is faced with two fundamental dilemmas. The first is that of going forward 'at two speeds', with some members leading and others lagging behind. Unless measures are taken to redistribute the benefits of the single market in a significant way there is likely to be an increasing gap between the EC 'haves' and the 'have nots'. The six original members seem to occupy the strongest position. The newer members may find themselves left behind (Smith and Wanke 1993). The second is that of 'deepening' versus 'widening'. This refers to the relative priority that is given to increasing the level of integration among existing members as opposed to expanding membership. The problems with implementing the Maastricht Treaty probably mean that the latter strategy will prevail, at least in the short-term. First the former EFTA countries and then the three Eastern European countries (and, perhaps, Turkey) will be admitted. The Maastricht design will not fade away, however, as long as the French and German governments see a deepening of the EC as a means of assuaging their fears of one another.

Multinational business has certainly hedged its bets on the outcome of European unification. Although some multinational firms are edging towards being organized on a regional or bloc basis, for multinationals as a whole in Europe this trend has not proceeded very far. For example, the sales of US manufacturing affiliates in Europe are still fairly evenly split between the national markets in which they are located and export markets. The biggest national markets – Germany, France, Italy and Britain – are particularly dominated by national affiliates. There is little evidence here, therefore, of multinationals organizing for the EC as a whole (Thomsen and Woolcock 1993). If the EC is a 'fortress', it is one filled with holes.

The other two nodes in the triad are much less developed than the EC. The NAFTA was only negotiated in 1992 (see, e.g., Randall 1992). In contrast to the EC which always insists on political reform prior to membership, Mexico was included without any requirement of reform in Mexican politics and bureaucracy. Opponents of NAFTA in the United States point to the lack of enforcement of labour laws and environmental standards in Mexico as potential incentives for the movement of jobs from the US to Mexico (Faux 1993). This is without noting the depth of the economic gradient which separates the US and Mexico. Proponents argue that in the long term NAFTA will raise incomes in Mexico and thus discourage migration to the US. The three countries are already trading partners, so the trade-diverting effects are not likely to be very great. Of course this is more important for Canada and Mexico than it is for the US. Given the number and relative power of the participants, it seems clear that while NAFTA will have relatively minor effects on the US economy as a whole it will give the US the ability to present itself as the centre of an American entity which extends beyond the borders of the United States.

Japan has no formally announced scheme to tie its economy more closely to East Asia's; yet links of trade and investment are multiplying. The historical links of Japan to the region were substantial prior to the Second World War but subsided afterwards (Cumings 1984). Since the 1970s the Japanese government has used its foreign-aid programme to build infrastructure in East Asia which then benefits Japanese firms. In the 1980s Japan replaced the United States as the largest source of FDI in the region. Much of this investment reflects the lower wages manufacturing firms need to pay in the rest of Asia compared to Japan, though some is based on the growth of final markets for goods. The yen has emerged to challenge the use of the dollar in transactions throughout the region. Japanese firms are also of rising importance in the region's financial markets (Frankel 1991).

Only by a considerable stretch of the imagination can this Japanese influence be characterized as a trading bloc, either actual or incipient. In the 1980s within-region trade in East Asia grew more slowly than did the region's trade with the rest of the world. Japanese FDI in East Asia during the same period did rise, but it was still at levels well below that of Japanese FDI in the EC or the US (Frankel 1991). By American or European standards there is still little intra-industry trade between Japan and its Asian neighbours. This will restrict the use of the yen (Ito 1992). Finally, memories of the Second World War are still very much alive in East Asia (Grant 1993). Because of its history in the region, Japan is often reluctant to take on any appearance of coercing its neighbours into following its policies. This is likely to limit the development of a formal entity such as the EC in East Asia so long as it depends on Japanese initiative.

The bottom line on trading blocs as the geopolitical order of the future, therefore, is that there is only one at present, and it is not doing very well. Threats of trade wars between the US and the EC or Japan over tariffs and non-tariff barriers to trade certainly can be interpreted as 'struggles' in the 'triad'. But this is to take considerable liberty with the concept of a trading bloc. At present over one-half of world trade is conducted outside of the triad. Trends towards the increase of international exchange (trade and investment) in general outweigh the trend towards regionalization. One of the most thorough studies of 'polarization' in world trade (Lloyd 1992, 35) concludes that the evidence for a growing regionalization of world trade is weak:

> The shares of RTAs [Regional Trade Arrangements] in total world trade have increased but this is explained in large part by the increase in membership of the RTAs and in any case it is a measure of the importance of these countries in *world trade* rather than of regionalization. (emphasis added)

CONCLUSION

The theme of this chapter has been the 'pursuit of primacy' and the qualifications of various actors to fulfil this role. In our view a geopolitical order can exist without the pursuit of primacy as a motivating force for its Great Powers. However, accepting for purposes of exposition that this is not the case, the possible accession of a set of candidates or 'hegemonic' pretenders has been reviewed. No particular country or trading bloc provides striking evidence for its incipient emergence as a successor to the United States. Rather, the burden of the evidence is that the world economy and the states within it are no longer conducive to the appearance of a singular state-territorial hegemon. This conclusion also applies to the idea of the 'triad' of 'trading blocs' that in some accounts is seen as the dominant spatial form of the 1990s.

The next chapter provides an alternative perspective on the geopolitics of the contemporary international political economy. This focuses on the twin processes of globalization and fragmentation alluded to briefly in Chapters 1 and 4, but it also addresses the fate of the state, the dominant geopolitical discourse of the time (transnational liberalism), and the changing geography of global development.

7

TRANSNATIONAL LIBERALISM

In the past two chapters we have charted the reconfiguration of US hegemony and the absence of any real pretenders to the US throne among Germany, Japan, China, Russia and the EC. In this chapter we return to questions of hegemony, hierarchy, order and territory from another angle. The focus here is on the rapid internationalization of economic and political affairs in the period since 1945, and the dangers and opportunities that might be presented in the realm of geopolitics by a *decentring* and *deterritorialization* of the means of production, destruction and communications. We join with some other commentators in linking a tentative globalization of modern life to the construction of overlapping sovereignties and networks of power that are in turn associated with a new form of hegemony: what we shall call *transnational liberalism*.

To speak in these terms is to insist on a Gramscian account of hegemony. We do not believe that 'After Hegemony' (Keohane 1984) there is no hegemony simply because no one nation-state is in a position to play the part of hegemon. It is rather the case that a new ideology *of the market* (and of market-access) is being embedded in and reproduced by a powerful constituency of liberal states, international institutions, and what might be called the 'circuits of capital' themselves. Whether this new hegemony can hope to be stable – and if so how, and in what terms – is a moot point, and one that we return to at the end of the chapter. It is also not yet clear how a new ideology of transnational liberalism might be associated with particular conceptions of hierarchy and territory. We offer some thoughts on this matter at various points in this chapter and more systematically in Chapter 8.

The rest of the chapter is organized as follows. The second part of the chapter provides an account of internationalization and globalization in respect of production, finance, telecommunications and information technologies, the means of destruction, and the environment. Our coverage is necessarily partial, but we are able to document not only a secular rise in the internationalization of economic and political affairs, but also a qualitative shift in the way these affairs are now organized. It is always easy to believe that today's times are 'new times', and recent theories of post-industrialism,

164

post-Fordism and post-modernism all suffer from a desire to embrace change over continuity. Nevertheless, recent developments in financial markets and information technologies have changed the ways in which peoples, places and 'countries' (or internationalized states) interact *and* how economic and political actors perceive these interactions. The proliferation of nuclear weapons since 1950 has had a not dissimilar effect on the possibilities of extended warfare between certain combatants. The new world order cannot be the same as a previous world order.

The third part of the chapter begins with some suitably pessimistic comments on the corrosive effects that 'globalization' has had (and may yet be having) on *international* order in a traditional IR-sense. It is not difficult to link the recent crises of inflation, debts and deficits to a generalized condition of hegemonic instability. The main thrust of this section, however is to outline the contours of an emerging hegemonic order of transnational liberalism. This section is thus concerned with the internationalization of modern states and with the renewed importance of international institutions like the IMF, the World Bank and GATT in policing a new regime of market-access economics and geopolitics. A subsequent section continues this analysis, but with reference to a new ideology of market-access economics and the activities of leading market-makers like pension funds, insurance companies and multinational corporations. The final section of the chapter reflects on the significance of transnational liberalism for traditional IR accounts of order, hegemony and territorial hierarchies. It also points up the Janus-faced nature of the new hegemony: of an ideology of market-access that at once embraces great asymmetries in power and wealth, and that is often willing to see the poor and 'imprudent' punished in support of some alleged greater good, but which also contains within it a vocabulary and a technology that can create radically new possibilities for empowerment and democratization. We consider this theme further in Chapter 8.

INTERNATIONALIZATION AND GLOBALIZATION

The world system mapped out at Bretton Woods and Yalta in 1944/45 held a three-fold significance for the mainstream international relations community: it served, first, as a *de facto* statement of international affairs, with the world system seeming to divide neatly into three worlds at the pinnacle of which stood the USA; it served, second, as a map of international relations, in which solid nation-states (rather like billiard balls) approached one another in areas as diverse as trade and warfare without suffering an unintentional loss of sovereignty; and it served, third, as a model of how stable international affairs could and should be organized. At least from about 1950 to the late 1960s, the Cold War Geopolitical Order coincided with a golden age of capitalism in the advanced industrial world (Marglin and Schor 1990), and with a spirit of optimism in much of the Second and Third Worlds. The international relations community was encouraged to believe that this was how the world system should be ordered (via a clear hierarchy of states), and to worry about possible threats to US hegemony and so-called balances of power and terror. These worries amplified greatly in the 1970s and 1980s as the world economy itself ran into trouble (Brett 1985).

In retrospect, it is clear that the very success of the post-war settlement also set in motion certain processes that would undermine this vision of international relations: a fixed hierarchy of autonomous nation-states enter-ing into relations with one another according to certain rules of conduct established and monitored by the USA (and the USSR). Although Maynard Keynes, as a signatory of the Bretton Woods agreement, looked forward to a post-war expansion of trade between independent nation-states, he did not anticipate that capital flows between porous 'nation-states' would come to serve as the most vital signifiers of the modern world economy (Moggridge 1992). Yet the relative stability of the post-war years helped to create a set of economic and political conditions in which capital would be international-ized not only in the form of commodity trade, but also through portfolio and direct foreign investment in production and allied service industries. The rapid development of new communications and information technologies has been part cause and part consequence of this tentative trend to globalization in economic affairs; a trend that also holds implications for warfare and ecological relationships.

The globalization of production:
TNCs, NICs and a new international division of labour

It is perhaps not surprising that economic and political commentators divided the world into three parts at the end of the Second World War. Although the territorial US economy was home to about 60 per cent of world manu-facturing production in 1945, it was apparent to the architects of the *Pax Americana* that a ruralized Germany (and possibly Japan) would serve no useful purpose for an outward-looking America and, indeed, that recovery and industrial reconstruction were bound to succeed in Western Europe and Japan in the medium term (Wachtel 1986). These countries had collective memories of industrialization and the necessary technical skills to rebuild strong national economies. A First World of industrial powers would soon extend beyond the borders of the United States. Meanwhile, the Second World was defined at first by the Soviet Union and its client states in Eastern Europe, and was later widened to include the new Communist states of China and Cuba. The success of nationalist–Communist revolutions in the last two countries muddied the waters a bit in terms of an emerging equation between a Second World and a bloc of socialist industrializing countries, but this local difficulty was not considered too disturbing. For its part, the Third World was a residual grouping of countries that included the present and ex-colonies of the West European powers, and was marked out for economic and political modernization or development. The Third World was also a fertile area in which the First and Second Worlds could engage in struggles for 'hearts and minds', as well as in less glorious proxy wars (Pletsch 1981).

Fifty years later this simple classification of countries has lost most of the validity it might once have had. Only China remains as a large socialist country in the Second/(Third) World, and even then it is socialist mainly in name and with regard to certain forms of political management. The Chinese economy is rapidly being liberalized and new industrial sectors are growing up under private ownership. The collective farms are largely a thing of the past (Nolan 1988). The non-socialist Third World, meanwhile, has fallen apart. Although large parts of the ex-colonial world continue to be haunted by absolute poverty (about 40 per cent of the population in South Asia: Singh 1990), by famines and food shortages (in parts of sub-Saharan Africa: Watts 1989), and by active underdevelopment (if one compares per capita incomes in parts of Latin America in 1990 with incomes in 1980: see also Scheper-Hughes 1992), other countries and city regions have prospered beyond all expectations. South Korea and Taiwan each recorded average annual rates of growth of GNP and exports in excess of over 9 per cent and 20 per cent respectively throughout the period 1965–90, and both countries are now counted as newly industrialized or upper-middle-income countries (Amsden 1989). Other countries that are talked of in not dissimilar terms include Singapore, Malaysia, Hong Kong, Bahrain and even, though less convincingly, Mexico and Chile. In recognition of these and other developments the World Bank now refuses to talk of First, Second and Third Worlds, but rather classifies countries as Low-Income, Middle-Income, Upper-Middle-Income and High-Income Economies (World Bank 1993).

What the World Bank has not yet recognized, however, along with the mainstream international relations community, is that islands of 'First World' activity are now present in many low-income economies, including Nigeria and India. By any standards India is still one of the poorest 'countries' in the world, even if its 1993 World Bank ranking as the world's 19th poorest country is rendered less than significant by a refusal to compare countries in terms of purchasing-power parities. At the same time, though, India has a middle class that numbers about 150 million men and women and that has become raucously consumerist since Prime Minister Narasimha Rao and Finance Minister Manmohan Singh set about liberalizing the Indian economy in the early 1990s. Delhi has never been an archetypal 'Third World city', but the divisions between Old and New Delhi are now more glaring than ever. Parts of south New Delhi are at least as prosperous as the poorer boroughs of London and New York, and the downtown area of the city is dominated by buildings that are home to such familiar global corporations as Citibank and American Express. Third World cities have long been described as 'shared spaces' (Santos 1979), but this term is now more apposite than ever.

City regions are emerging in countries like India, Brazil and Nigeria that are as much parts of an internationalized 'First World' as they are parts of the (ex-)Third World (King 1990). Against such a background it makes little sense to map out the world economy in terms of a ranked hierarchy of countries running from Switzerland to Mozambique (as per the World Bank's *World Development Report* for 1993). One could equally refer to a global hierarchy of city regions or central places that links a first tier of world cities – including London, Tokyo and New York – to second, third, fourth and fifth orders of cities, the latter of which would include parts of New Delhi, Rio and Lagos. One of the important recent texts in international relations is entitled *The Hierarchy of States* (Clark 1989). In the 1990s it will not suffice to refer to this hierarchy alone; states are not undivided, nor are they unchallenged by other economic and political actors (Camilleri and Falk 1992).

But what has brought about this partial reordering of global space? When *deterritorialization* first became a topic for discussion in the 1970s attention was largely focused on the activities of multinational (or transnational) corporations in the forging of a new international division of labour (Murray 1971). TNCs existed before 1945, of course, but their rate of formation increased significantly after 1950. An estimated 55 overseas subsidiaries of American corporations were formed annually between 1946 and 1952, a doubling of the average rate recorded in the inter-war years. The rate doubled again between 1953 and 1955, until more than 300 new subsidiaries were being set up annually in the years between 1959 and 1962. The origins, or headquarters, of TNCs also changed over this period, with an early domination of US and UK-based companies giving way slowly to a more even distribution of US, UK, German, Dutch, French and Japanese-based companies among the world's top 500 corporations (Dicken 1992).

The economic conditions that allowed TNCs to prosper were bound up with the economic and political settlements that ushered in the Cold War Geopolitical Order. According to Stephen Hymer, we might say that a first wave of US-based multinationals emerged in response to the particular conditions of the 1950s: conditions that included market saturation and oligopolization in the USA; an increasingly sophisticated international communications and transportation network; an evolving hierarchical division of labour; and a growing economic challenge to the USA from Europe and Japan. It was against this background that leading US corporations decided upon an 'outward thrust to establish sales production and bases in foreign territories' (Hymer 1975, 47) and most notably in the UK and the EC after its inception in 1957. More generally, large trading corporations were encouraged to move abroad by the emergence of offshore financial markets (see next sub-section), by declining rates of profit in some metropolitan countries, and by the possibilities of exploiting new markets and supplies of labour in some less developed countries (Thrift and Taylor 1989).

In terms of the most favoured locations for TNC investment, there is no evidence to suggest that these modern Leviathans have preferred authoritarian countries to democratic countries or vice-versa (Sheahan 1980). But some US companies did locate at first in countries with strong geopolitical links with the TNC's host country (such as Mexico and South Korea), just as UK and French-based multinationals have maintained a presence in many ex-British and French colonies. TNCs have also been encouraged to move to those countries that have maintained a welcoming attitude to direct foreign investment. It is no surprise to learn that India was not a favoured location for TNCs until the country embarked upon the recent liberalization programmes pushed through by Manmohan Singh; TNCs preferred the tax breaks offered to them by less economically nationalistic countries like Taiwan and (after 1973) Chile. Finally, it is worth noting that TNCs have been attracted to developing countries that are reasonably well managed and that show signs of economic growth and development. TNCs do not generally set up shop in countries like Bangladesh or Sudan, no matter how low unit labour costs might be in such countries. They prefer to locate in countries where labour is cheap by the standards of the high-income economies, but which is also skilled or semi-skilled and disciplined in the ways of the market and time-management (Porter 1990). TNCs also have a preference for locating in countries with stable governments and in countries where the domestic market is growing rapidly, or at least is not stagnant.

The internationalization of production capital has several observable consequences for concrete *spatial practices* and our *representations of space*. On the one hand, it is now widely accepted that today's giant trading corporations have a global reach that is comparable to that of many small and middle-sized nation-states. In the early 1980s it was common to find tables in textbooks akin to that reproduced here as Table 7.1; tables that ranked TNCs and countries together in terms of the sales turnovers of the former and the GDPs of the latter. By such means it was possible to suggest that a company the size of Exxon, Coca-Cola, or the IBM of old was as sizeable a global player as a country like New Zealand or Pakistan, not to mention Czechoslovakia (as was) or Bolivia. The implicit argument was a flawed one because size alone is a poor proxy for political power, but in other respects such a comparison did make a significant point: namely, that nation-states were not the only sizeable actors in the global political economy of the post-1970s period. Rather more telling arguments also pointed out that about 40 per cent of world trade in the early 1980s, as in the early 1990s, was in the form of trade between branches of the same multinational company (intra-firm trade: Grimwade 1989), and that multinationals were promoting a globalization, or Americanization, of culture that found its zenith or nadir in the transmission of the television soap-opera *Dallas* to 106 different countries in the late 1980s. To the extent that hegemony in a geopolitical sense is bound up with cultural imperialism, this is arguably a telling statistic (Peet 1989; Wallerstein 1992).

Table 7.1 The world's top twenty corporations, 1992 (by market value) and selected countries, 1991 (by GDP)

Rank	Name of corporation/country	Market value/GDP Billions of US dollars
1	Royal Dutch/Shell Group (Neths/UK)	77.82
2	Nippon Telegraph & Telephone (Japan)	77.52
3	Exxon (US)	75.30
4	Philip Morris (US)	71.29
5	General Electric (US)	66.00
6	Wal-Mart Stores (US)	60.82
7	Coca-Cola (US)	58.47
8	Merck (US)	58.41
9	AT&T (US)	55.85
10	IBM (US)	51.82
11	Toyota Motor (Japan)	43.97
a	*New Zealand*	42.86
12	Glaxo Holdings (UK)	42.64
13	British Telecom (UK)	40.45
b	*Pakistan*	40.24
14	Mitsubishi Bank (Japan)	39.84
15	Bristol-Myers Squibb (US)	37.60
16	Sumitomo Bank (Japan)	37.12
17	Du Pont (US)	35.41
18	Proctor & Gamble (US)	34.74
19	Dai-Ichi Kangyo Bank (Japan)	34.68
20	Industrial Bank of Japan (Japan)	34.04
c	*Czechoslovakia*	33.17
d	*Bolivia*	5.02

Source: *Business Week*, 13 July 1992; World Bank 1993, Table 3

Above and beyond these events and observations, the internationalization of production capital has been significant for the way that it has helped to explode the unity so often assumed to exist between a given country's 'economic interests' and the interests of a particular territorial economy. We have documented this already in the case of the USA (see Chapter 5), but such a geographical mismatch is also apparent in the cases of the UK and the Netherlands, and increasingly in Germany and Japan. All leading industrial countries now rely heavily on foreign direct investment, and in some of these countries long-established 'foreign' firms are now defended as domestic businesses. (We think here of Ford and General Motors in the UK, both of which were believed to be under threat in the 1980s from a new wave of foreign – for which read Japanese – inward investment in the 'UK' automobile industry.) The fact that some companies are assimilated in this way is very significant. It reveals both the sedimentation of 'foreign firms' in supposedly alien territorial environments and the real difficulty that some economic and political actors – institutions as well as individuals – have in separating out domestic and foreign employers. The fact that the division itself is of decreasing consequence, and that 'foreign' firms cannot easily leave some territorial economies without writing off expensive fixed capital investments, is surely also suggestive of certain changes in the modern geopolitical economy (Knox and Agnew 1994).

In the 1930s and 1940s it was possible for countries in Europe and elsewhere to go to war with one another to defend singular national economies and local alliances of national capitals and a national labour force. This is no longer an easy option. The internationalization of productive assets raises interesting questions about the nature of modern economic sovereignties and about the costs and benefits of warfare between economically interdependent countries. The dramatic post-war growth of international trade between the core 'Fordist' countries has reinforced this growing sense of caution. (Between 1950 and 1990 trade between the high-income economies grew by a factor of 15.) By 1991, even hitherto 'closed' economies like the US and Japan had import/GDP ratios of about 7–9 per cent, while more open economies like the UK and the Netherlands (not to mention Taiwan and South Korea) had ratios closer to 25–45 per cent (World Bank 1993). With 'national' prosperity depending so much on the prosperity and stability of a given country's principal trading partners, it is not surprising that the prospect of conventional wars between high-income countries has diminished sharply over the past 50 years. Even the threat of protectionism is not all that it might seem. It is possible, of course, that the high-income economies will seek to impose further tariff and non-tariff restrictions on imports from the industrializing regions of the ex-Third World, but protectionism between the advanced industrial countries courts a very real threat of economic retaliation (Gill and Law 1988). Protectionism also runs counter to an emerging ideology of transnational liberalism that stresses the efficiency and equity of voluntary economic exchanges in the global market-place. Protectionism has a constituency in all of the most prosperous countries in the world and it is used as a threat with monotonous regularity; in practice, however, protectionism is not yet firmly on the agendas of the high-income economies in their dealings with one another, although it is often used against poorer countries seeking to join the rich world's 'club'.

Finance

The activities of most TNCs are rightly dealt with under the rubric of the internationalization of capital. Although these corporations are happy to engage in transfer pricing in order to maximize their after-tax profits on a world-wide basis, they are none the less firmly headquartered in one country and firmly anchored in some other countries by virtue of the investments they have made there. Globalization, properly speaking, should mean something more than this, as, for example, when individuals in different countries are able to consume images simultaneously by virtue of satellite TV communications. In the economic realm globalization is probably most advanced in the service industry sector, and particularly in the area of financial services. Recent developments in this industry have had dramatic implications for the nature and extent of transactions between spatial actors – including so-called nation-states.

One way to see that this is so is to step backwards towards 1940 to bring the story forwards into the 1980s and 1990s. The Bretton Woods agreement of 1944 led to the setting up not only of the International Monetary Fund (IMF) and the IBRD (the International Bank for Reconstruction and Development, now part of the World Bank Group), but also an international monetary system based around a gold–dollar exchange rate standard and fixed (or fixed-peg) exchange rates. The main international unit of account was to be the US dollar, but the value of the greenback was to be defined in terms of gold ($35 for one ounce) in order to dissuade the US government from printing dollars to excess (Corbridge 1994a). The US government also worked with its allies in Western Europe and elsewhere to guarantee a political and economic climate that would pave the way for currency convertability in the late 1950s.

As Robert Gilpin has pointed out (following John Ruggie):

> [The] Bretton Woods system was a compromise solution to the conflict between domestic autonomy and international norms. It attempted to avoid (1) subordination of domestic economic activities to the stability of the exchange rate embedded in the classical gold standard and also (2) the sacrifice of international stability to the domestic policy autonomy characteristic of the inter-war period. This so-called 'compromise of embedded liberalism' was an attempt to enable governments [in the First World] to pursue Keynesian growth stimulation policies at home without disrupting international monetary stability.
>
> (Gilpin 1987, 132)

The problem was that this compromise could not be sustained. The Bretton Woods system both embodied US hegemony in international economic and financial relations and depended upon this hegemony. If and when the economic and financial powers of the USA began to be challenged by rival economic actors, and if and when the USA began to abuse the powers of financial seigniorage that had been accorded to it as the issuer of the world's reserve currency, the system would break down.

The weaknesses of the Bretton Woods system were probably first highlighted by Professor Robert Triffin, an economist at Yale University, in his book *Gold and the Dollar Crisis* (Triffin 1960). Triffin pointed out that:

> a fundamental contradiction existed between the mechanism of liquidity creation and international confidence in the system ... Either America's balance-of-payments deficits had to stop (thereby decreasing

the rate of liquidity creation and slowing world economic growth) or a new liquidity-creating mechanism had to be found.

(as summarized by Gilpin 1987, 135)

The so-called Triffin Paradox became more pressing in the late 1960s and early 1970s when the US began to pay for its adventures in Vietnam by printing dollars to excess and running up severe balance of payments deficits (Wood 1986). At this time, and in this manner, the US was exploiting the first of two basic asymmetries that eventually destroyed the Bretton Woods system in 1971–3. First, the US was able to conduct its foreign policy more or less in abeyance of concerns about its balance of payments position, because it issued the very currency that it offered to other countries as IOUs. The US attended to a second asymmetry in the system in August 1971 when President Nixon abrogated a previous commitment on the part of the US to maintain the US dollar at a fixed value in terms of gold. Henceforth the USA would be able – and in practice was more than willing – to devalue the dollar in order to improve the trading position of the US economy (Parboni 1984; 1988).

Seen in these terms it should be apparent that the breakdown of a system of fixed exchange rates in the early 1970s was largely a response to the changing position of the US in the world economy, and of the unwillingness of some European countries meekly to accede to a revaluation upwards of their own currencies. As we noted before, it was the very success of the *Pax Americana* (the Bretton Woods system, Marshall Aid, Cold War spending, and increasing international flows of production capital) that undermined the economic hegemony of the US territorial economy relative to the territorial economies of some of its rivals in Western Europe and East Asia. But none of this means that institutional and technical developments in the fields of international money and finance were not also important in the restructuring of international financial and geopolitical relations; they were.

The Bretton Woods system was not only a system that embodied national controls on exchange rates (subject to the stop–go constraints of balance of payments concerns outside the USA); it also enshrined the principle of national sovereignty in the production and transmission of paper monies (Eichengreen and Lindert 1989). With the development of the Euromarkets in the 1950s and 1960s this principle of monetary sovereignty gradually was eroded. The Euromarkets probably came into being in the early years of the Cold War period, when the governments of the USSR and China began to deposit some of their US dollar holdings in banks that were beyond the regulatory reach of a hostile US administration. Given that many of these banks were located in London and other European cities it became a convenience to refer to a burgeoning Eurodollar market, although this same term today would refer to all currencies deposited outside the borders of the country in which they were printed and first issued. Alain Lipietz has suggested that contemporary offshore markets are trading in 'xenocurrencies'

(Lipietz 1987), and his wording certainly has a good deal to commend it. Howsoever we refer to these markets, it is clear that they expanded massively from about 1960, and in the capitalist world economy rather than in the socialist bloc. The first masters (and servants) of the Eurodollar markets were US multinationals in Europe. As Richard O'Brien points out:

> In the early days the demand for eurocurrency loans came as a result of an increase in activities in Europe by U.S. multinationals. One of the first borrowers of eurodollars was IBM Europe. Although this was in the days of fixed exchange rates and thus limited FX [foreign exchange] risk, lending dollars to a company in Europe might have been considered problematic if that company had no clear stream of dollar earnings or access to dollars for repayment. IBM Europe, however, was seen as ultimately having access to dollars through its parent, and thus lending to IBM Europe did not pose any 'geographical' mismatch problems.
>
> (O'Brien 1992, 30)

The Eurocurrency markets have developed a long way since the early 1960s – so much so, indeed, that geographical mismatch problems are now rarely taken seriously. The markets expanded considerably in the 1970s on the basis, first, of the monetization of the US deficits following the breakdown of the Bretton Woods system and the escalation of the war in Vietnam, and, second, following the OPEC oil price rises of 1973–4. The West's commercial banks were now called on to recycle the Middle East's petrodollars through the London-centred Euromarkets to a group of oil-importing middle-income countries in Latin America and South-East Asia. To meet this obligation the leading money-centre banks themselves internationalized their activities and began to issue new financial products, including the syndicated bank loan that was (re-)invented by Walter Wriston's Citibank. By the mid-1970s about 70 per cent of Citibank's overall earnings came from its international operations, with Brazil alone accounting for 13 per cent of the bank's earnings in 1976 (Makin 1984, 133–4), and various island micro-states, including the Caymans and the Bahamas, were consolidating their positions as offshore banking centres (OBCs) booking loans to and from different centres in the two Americas (and elsewhere). In the mid- and late 1980s many of these OBCs became more broadly-based offshore financial centres (OFCs), with many making a tidy profit on serving the tax-avoidance 'needs' of a growing global community of seriously wealthy individuals (see Figure 7.1).

The emergence of fictitious spaces trading in fictitious capitals (Roberts 1994) is emblematic of certain wider developments in international financial relations in the 1980s. One such development concerns the nature of money itself and of various widely traded financial products and derivatives. At the turn of the twentieth century money still had a tangible quality to it in the form of the gold and silver reserves that closely backed the paper or fiat monies issued by governments. In a sense, this tangibility of money was

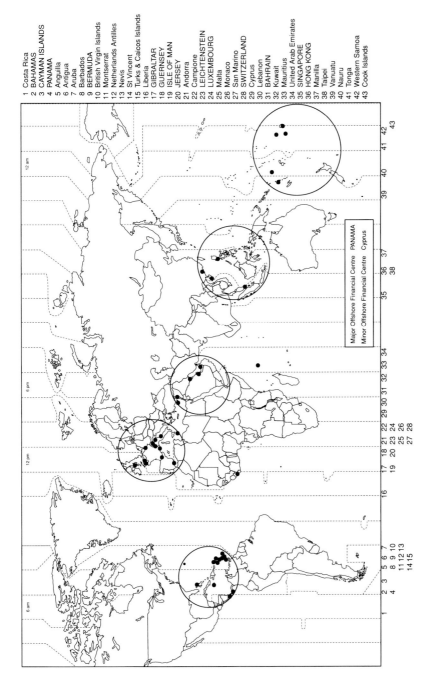

1 Costa Rica
2 BAHAMAS
3 CAYMAN ISLANDS
4 PANAMA
5 Anguilla
6 Antigua
7 Aruba
8 Barbados
9 BERMUDA
10 British Virgin Islands
11 Montserrat
12 Netherlands Antilles
13 Nevis
14 St Vincent
15 Turks & Caicos Islands
16 Liberia
17 GIBRALTAR
18 GUERNSEY
19 ISLE OF MAN
20 JERSEY
21 Andorra
22 Campione
23 LEICHTENSTEIN
24 LUXEMBOURG
25 Malta
26 Monaco
27 San Marino
28 SWITZERLAND
29 Cyprus
30 Lebanon
31 BAHRAIN
32 Kuwait
33 Mauritius
34 United Arab Emirates
35 SINGAPORE
36 HONG KONG
37 Manilla
38 Taipei
39 Vanuatu
40 Nauru
41 Tonga
42 Western Samoa
43 Cook Islands

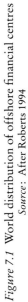

Major Offshore Financial Centre   PANAMA
Minor Offshore Financial Centre   Cyprus

Figure 7.1 World distribution of offshore financial centres
Source: After Roberts 1994

reinvented in 1944 when the dollar was made as 'good as gold'. Since 1971, however, the international monetary system has had to depend on a paper money standard pure and simple (augmented by such amalgam currencies as the IMF's Special Drawing Right and the EC's European Currency Unit). Money has never been an easy commodity for individuals to understand, a point that Dickens understood very well when he had Dombey junior say to his father 'Papa, what's money?' (after Harvey 1989, 167). In the contemporary world economy, however, the funny and even devilish qualities of money are more apparent than ever (Watts 1994). People are now required to place their faith – and the products of their labours – not only in commodity and paper monies, but in electronic representations of these paper tokens of 'real money' (Rotman 1989). They are also expected to place their faith – and in some cases their savings – in financial derivatives like securitized mortgages, futures contracts and Eurobonds, and in the financial institutions and markets that largely guarantee their worth (Corbridge and Thrift 1994).

It is this last point that holds a particular significance for geopolitics. On the one hand, the denationalization and deterritorialization of liquidity creation and transmission clearly has implications for IR models that place a premium on the economic sovereignty that supposedly is embodied in nation-states. In today's money markets, the 'hard money' put into circulation by governments and central banks is being expanded according to the activities of commercial bankers who answer mainly to the international financial markets. Although the monetary multiplier of the Eurocurrency market does not in practice approach infinity – that is, commercial bankers will not make loans (expand their asset base) entirely without regard to the credit rating of a particular client – the lending activities of some commercial banks in the 1970s and early 1980s, and again during the property boom in some Western countries in the late 1980s, does point up both the power of some banks to expand the global money supply, and the increased capacity for 'dis-coordination' in 'national money supply policies' (Lissakers 1991). It is not clear how states can effectively control national or international money supplies in the absence of clear and coordinated information on the scale of money-supply activities in the private financial sector. (The Bank for International Settlements has recently issued new capital adequacy guidelines in an attempt to deal with this problem; see also Leyshon 1994.)

In still broader terms, it is clear that the financial services industry is showing signs of an incipient globalization of some of its products and activities. It remains true, of course, that US banks tend to play a dominant role in Latin America, just as Japanese financial institutions are entrenched in South-East Asia and German banks are involved in the reconstruction of central and Eastern Europe (Leyshon 1994; Marsh 1992). Banks still need to maintain a physical presence in different countries and the expense of this cautions most commercial banks against a truly global presence. When it comes to money itself, however, international boundaries are becoming less and less relevant. As O'Brien points out: 'the euromarket offered a global pooling of funds which could be tapped by borrowers from anywhere (as long as their own domestic authorities permitted the borrowing of foreign currencies) and where depositors could place their monies (again subject to national restrictions)' (O'Brien 1992, 33). Whereas international trade commonly involves the physical movement of one commodity from one country to another, the fungibility of money in a deregulated, post big-bang world ensures that even customs posts and national identities are often of little monetary significance. The financial exchange (FX) market is a good example of this. Whilst it is true that the FX market is international in the sense that 'all exchange rates involve different countries' (O'Brien 1992, 34), it is globalized in the sense that all currencies are exchanged in one way or another, in that there are few entry barriers to the FX market, in that the FX market is a 24-hour market, and in that the FX market has no special home, but is played out on computer screens and over the telephone. Much the same holds true of some financial derivatives, including swaps and futures contracts and markets. It is also worth noting that the securities industry has undergone a similar internationalization since the mid-1980s. In Amsterdam, at the end of 1990, 47.8 per cent of all companies quoted on the local Stock Exchange were foreign-based. In London, 553 foreign companies were listed on the Stock Exchange at the same time (21.6 per cent of total listings). In London too, in 1990, 'the turnover of foreign equities exceeded that of domestic stocks for the first time, primarily [as] a result of the growth of SEAQI (the exchange's screen-based market for trading foreign equities)' (O'Brien 1992, 44–5).

The significance of these changes in the global money market should not be underestimated. We have not yet reached the 'end of geography' that O'Brien seems to be looking forward to, albeit with tongue firmly in cheek, and far less have we reached the 'one world' beloved by all manner of Utopian thinkers and sundry idealists. The differences in income levels between rich and poor peoples and places continue to widen on a global basis (Arrighi 1991). Nevertheless, we do now live in a global economy that is thoroughly internationalized in its expanding core areas in terms of commodity trade and production capital, and which is globalizing apace in the case of money flows and financial capital (Warf 1989). We also live in a world in which *the markets* can defeat even the most concerted efforts by a government, or even groups of governments, to defend particular national exchange rates and interest rates. The collapse of the European Community's Exchange Rate Mechanism in 1992–3 offers ample testimony to the powers of today's market-makers and to the poverty of national currency reserves when faced with massive money movements on the open market. The blunt truth is that money today, more than ever, is a type of information that is traded on computer screens and that refuses all but the simplest of national identities.

Despite early indications to the contrary (Feldstein and Horioka 1980), it is also likely that present-day correlations between domestic savings rates and rates of investment are low (see Cooper 1991, and consider the funding of the twin US deficits since the mid-1980s). The removal of exchange controls in the 1970s and 1980s (and not just in the high-income economies), together with the financial deregulations that defined the 1980s, mean that a global capital market exists now more clearly than at any time since 1914 and on a much expanded scale. Even powerful countries like the USA are having to adapt their economic policies (and presumably their geopolitical policies in a more conventional sense) to take account of the contours of an inter-nationalizing world economy in which real powers are rapidly draining away from singular countries or 'nation-states'. Whether these powers can be reclaimed is a moot point. In principle they clearly can, given the political will. In practice, given recent developments in information and communications technologies, it is not so easy to see how the genie can be put back in the bottle.

Information and communications technologies

According to Plato (and Marx!) necessity is the mother of invention, and there probably is something to be said for the view that most technical innovations are called forth by a wider set of economic and political determinants; technology should not itself be celebrated as a *deus ex machina*. This point accepted, what is significant about the modern geopolitical economy is the extent to which it revolves around the production and trading of information and communications technologies themselves, as well as being more broadly supported by them (Hepworth 1989). Developments in this field since 1950, and more especially since 1970, have radically changed the nature of global production and locational possibilities, and the boundedness of the 'nation-states' that are meant to house the economic engines of international politics (Moss 1987).

As Daniels points out, 'the two main advances [in this area] since the 1960s have been in microelectronics and in telecommunications' (Daniels 1993, 28). Both have had tangible effects on the internationalization of business and on the determinants of economic activity. In microelectronics the main break-through has been in terms of the storage capacity of a single silicon chip. The chip

> is the key component of everything from hand-held calculators through desktop personal computers to large main-frame computers and is [now] so reliable that the number of failures per unit of time has been reduced by a factor of at least 10 million since the 1950s.
>
> (ibid.)

One result of this breakthrough has been to reduce the size of computers and to improve their portability and levels of general use. Computers are now installed throughout the economies of the advanced industrial world and are being used increasingly in the private and public sectors of many developing countries. The bookings system of the Indian Railways system is now largely computerized, together with that of Indian Airlines; both industries significantly being devoted to the movement of people around a vast subcontinent and with the reduction of lost 'time as money'.

From the point of view of international relations, however, the development of micro-computers has probably been less significant than the coupling of microelectronic technologies to new systems of telecommunications. Daniels is again very informative on the topic. He notes that:

> For almost a century prior to the late 1940s telecommunications (the telephone and the telegraph) depended on analogue transmission. This gives direct and continuous wavelike representations of voice or audio signals; the speed of transmission is therefore slow and the kind of information transmittable is limited. Digital transmission involves the transfer of information which has been converted to binary form, leading to a marked increase in speed and capacity as well as in the diversity of information that can be handled. But this also has required an improved medium for transmission since the traditional copper-wire telephone networks are unreliable, slow and restricted in capacity. Fibre-optic cables (which use modulated light wave signals generated by lasers) have overcome these problems and have further underpinned the convergence of telecommunications and computing.
>
> (Daniels 1993, 28–9)

The emergence and rapid deployment of these linked technologies is leading both to an extraordinary escalation in the number of international connections of a traditional kind and to wholly new means of interfacing at-a-distance. Some idea of the scaling-up factor can be gained from an examination of the available statistics on international telephone and fax calls, airline flights and passenger movements, and migration figures, all of which bear testimony to particular forms of interdependence in the modern world.

Consider international telephone traffic in the first instance. The volume of international telephone traffic increased by more than ten fold in the 1980s and a similar rate of increase in the 1990s is forecast by the International Institute of Communications. Inevitably, most of this traffic continues to be between countries in the high-income world economy. In 1990 outgoing calls from the USA totalled 5.3 billion minutes; the equivalent figures from Taiwan and Ireland were 0.216 billion minutes and 0.075 billion minutes respectively (see also Table 7.2). But this 'telegeography' is changing rapidly in two respects. First, there is a general trend to increased use of telephone lines for non-traditional purposes so that, in 1990 again, 60 per cent of trans-Pacific telephone minutes involved the use of facsimile (fax) transmissions. This new pattern of usage is in turn being reinforced by the development of in-house telecommunications networks (including electronic mail facilities) in many

Table 7.2 International telephone traffic, selected countries, 1990

Country	Outgoing	Telecommunications traffic (million minutes) Incoming	Balance
Australia	518	398	−120
Japan	764	732	−32
South Korea	188	350	162
Malaysia	80	100	20
Taiwan	212	302	90
Austria	476	487	11
Belgium	731	755	24
Canada	565	358	−207
Denmark	362	343	−19
Finland	186	213	27
France	1,921	2,190	269
Germany	2,833	2,369	−464
Ireland	75	122	47
Italy	908	1,161	253
Israel	118	202	84
Luxembourg	151	83	−68
Netherlands	905	852	−53
Norway	281	277	−4
Portugal	126	270	144
Spain	611	653	42
Switzerland	1,356	1,016	−340
Turkey	159	441	282
UK	2,253	2,330	77
US	5,265	2,604	−2,661
Canada	565	358	−207

Source: International Institute of Communications 1991

of the large trading TNCs – many of which see it as a source of competitive advantage in a world where the information technology (IT) revolution 'has virtually eliminated the effect of distance on the time required to communicate an item of information between interacting locations' (Daniels 1993, 30). Second, the decreasing effectiveness of physical or legal barriers to communications and services transactions is rapidly extending to parts of the developing world.

Daniels' excellent account of the IT revolution was published in mid-1993 using data drawn mainly from 1988–90; at one point he claims that 'the unequal distribution of telecommunications services and traffic is underlined by the example of India, which not only lacks international connections but also has fewer telephones than London' (Daniels 1993, 31). By 1992–3 this statement was quite inaccurate. India was home to a veritable 'telephone revolution' in the early 1990s, mainly located in an efficient private sector, but working in tandem with a new government committed to economic liberalization. It is now possible to direct dial anywhere in the world from street corner kiosks in practically every medium- and large-sized town in India and STD lines are available for domestic callers. The government is also committed to the efficient delivery and installation of new telephones and telephone lines to subscribing businesses and households. The Chairman of the Telecom Commission, Nagarajan Vittal, spoke in August 1993 about his desire to reduce the waiting period for a new telephone connection from the two years it took on average in the late 1980s to two days by January 1996 (*India Today* 1993, 59). At the same time, the state-owned broadcasting service, Doordarshan, launched five satellite channels in 1993 to compete with the extraordinarily successful *Star TV* satellite system run out of the Gulf, India and Hong Kong by the Murdoch media and communications empire. India in the mid-1990s has a burgeoning middle class that is increasingly hooked into the global culture of consumerism advertised on the satellite channels and in the glossy magazines now being published in New Delhi and Bombay. Some commentators have linked a recent increase in the size of dowries in India (technically illegal but often of the order of $40,000 for an upper-middle income family) to the rampant consumerism that has been unleashed in India (as in East and South-East Asia) by the global communications media (Rao 1993). The global village that Marshall McLuhan anticipated in the early 1960s (McLuhan 1966) is becoming a reality for some households in the richer parts of Asia (and Latin America and Africa).

Indian business-people can also take advantage of a new range of transborder services linked to the IT revolution. Daniels and others draw attention to the new possibilities for transborder data flows in such goods as design proposals, share prices (via on-line information services like Quotron, Reuters and NASDAQ), currency movements, export manifests, loan agreements and so on. The companies that are providing services in this field, including British Telecom, Amex and Anderson Consulting, are among the fastest growing TNCs in today's global economy (see Table 7.3). These and other companies are opening up new opportunities for distance learning and for home shopping and banking. More so than most TNCs, these companies are instinctive globalizers and not just internationalizers. Such corporations both make possible and represent a new *world of flows* in an *informational society* (Castells 1988).

None of this means, of course, that people are now bound to stay at home or that goods are not traversing the world in ever greater quantities. We have already seen that international trade in commodities is fast expanding (albeit not as rapidly as international capital and financial flows), and further expansion is predicted in those regions of the world now embracing more outward-oriented development strategies; those regions now seeking access to the global market-place. Recent predictions suggest that international trade in the Pacific Basin will increase threefold in the 1990s. People are also on the move, both voluntarily and as a result of domestic compulsions. If international air passenger movements and destinations are a guide to voluntary

Table 7.3 Leading information service companies in the world, 1986 and 1989

Firm[a]	Home country	1986 Information services revenue[b]	1986 % of total revenue	1989 Information services revenue	1989 % of total revenue
Electronic Data Systems Corp.	US	0.0	0.0	2,447.9	45.3
Automatic Data Processing Inc.	US	1,298.1	100.0	1,689.5	100.0
TRW Inc.	US	1,450.0	24.0	1,565.0	21.3
Computer Science Corp.	US	977.7	100.0	1,442.8	100.0
Digital Equipment Corp.	US	0.0	0.0	1,386.7	10.7
Andersen Consulting	US	0.0	0.0	1,225.7	85.0
International Business Machines	US	300.0	0.6	1,200.0	1.9
Cap Gemini Sogeti	France	0.0	0.0	1,103.4	100.0
NTT Data Communications Corp.	Japan	577.6	1.8	898.7	39.9
Unisys Corp.	US	0.0	0.0	825.0	8.2
Black & Decker Corp.	US	0.0	0.0	687.6	21.6
American Express Co.	US	0.0	0.0	660.0	2.6
General Electric	US	550.0	1.5	550.0	1.0
Martin Marietta Corp.	US	0.0	0.0	502.2	8.7
Ernst & Young	US	0.0	0.0	450.0	11.5
NCR Corp.	US	350.0	7.2	425.0	7.1
STC PLC	UK	0.0	0.0	424.6	10.0
Sligos	France	0.0	0.0	385.5	96.2
SD-Scicon PLC	UK	0.0	0.0	381.3	82.3
British Telecommunications PLC	UK	0.0	0.0	360.2	1.8

[a] Ranked by total information services revenue, 1989 (millions of dollars)
[b] Value 0.0 denotes firms not operating in information services in 1986
Source: UNCTC 1991, after Datamation 1990 and 1987

movements, it is significant that virtually all the world's leading airports increased the number of destinations they served in the 1970s and 1980s, in most cases by an average of 40 per cent. London–New York City remains the busiest long-haul route for business passengers, but the region that saw the fastest rate of growth of air passenger movements in the 1980s and early 1990s was the Asia–Pacific region. Airports at Tokyo, Hong Kong, Los Angeles and Singapore are fast gaining ground on the traditional international hub airports of London, New York, Chicago, Paris and Frankfurt. Plans are also afoot for a new generation of wide-bodied jets that will carry close to 800 passengers instead of the present maximum of about 450. There has even been excited talk about a Japanese rocket plane. Although this plane is only at the design stage as yet, the intention is that the journey time from Tokyo to Los Angeles would be cut to just two hours. In the early 1990s the average commute-time in Tokyo was also two hours. Futurologists are not alone in envisaging a world in which Japanese business people might live in Tokyo and work in Los Angeles (or vice versa). Geography in the sense of distance is bound to matter less and less as new technologies come on stream to shrink space further, or to hasten what David Harvey has referred to as a process of time–space compression (Harvey 1990), and what Anthony Giddens calls time-space distanciation (Giddens 1984). In this regard, it is sobering to look again, or for the first time, at McHale's remarkable map of the world at different points of time wherein physical distances are represented in terms of average travelling times (see Figure 7.2).

Away from the world of the business-class traveller, many women are still having to spend four hours a day travelling ten miles to collect water and fuelwood (Agarwal 1986), and many more individuals are caught up in more or less forced movements from their home countries to those countries and regions willing to put them to work, often in degrading circumstances. It is estimated that 6 million slaves were taken from Africa between 1700 and 1810 to be put to work in the plantations, fields, houses and mines of the New World. The prosperity of large parts of the 'First World' depended on their labours (Wolf 1982). Today, the traffic in human labour is undiminished, even if it is open to at least some forms of international surveillance. Notwithstanding the fact that the hegemony of transnational liberalism, and market-access, does not extend to the free movement of labour (especially non-white labour), the economies of most high- and middle-income economies depend heavily upon legal and illegal flows of migrant labourers from poorer countries and regions. At the time of the Gulf War an estimated 600,000 migrant workers from South Asia were trapped in the region. In the USA, meanwhile, Spanish is rapidly becoming the second or even first language of some of the southern States, as Latinos join blacks, Asians and others in the fields and cities of California, Texas, New Mexico and Florida. The available evidence suggests that migrant workers everywhere contribute to enhanced rates of economic growth in the recipient country in the medium-term (Black 1991); sadly,

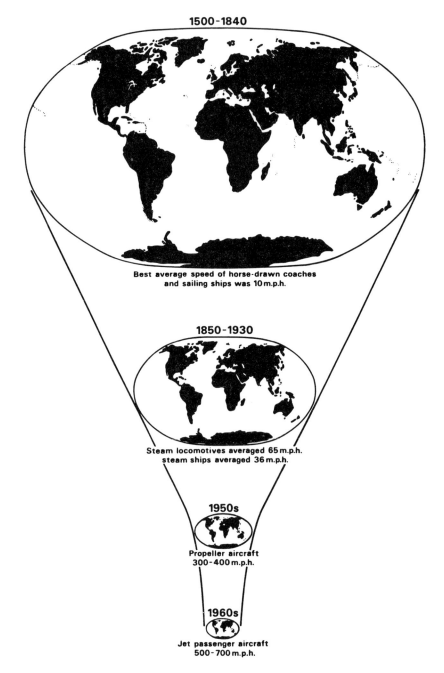

Figure 7.2 Global shrinkage: the effect of changing transport technologies on real distance

Source: Dicken 1992, 104 (following McHale 1969, Figure 1)

though, this has not discouraged the sorts of racist backlashes that are evident today in the UK, France and Germany, where 'immigrants' are often seen as unwelcome competitors for scarce jobs and housing. The internationalization of labour seemingly goes hand-in-hand with the reproduction of jingoistic and nationalistic ideologies and political parties. In the UK, 'Pakis' (a generic term of abuse linking together Pakistanis, Indians, Bangladeshis and others from South Asia) have become the latest in a long line of internal Others dating back most obviously to the Irish immigrants of the nineteenth century.

War and the environment

The dizzying interchange of information on a global basis, together with the uprooting and resettling of a growing population of international migrants, means that many political identities are no longer anchored in singular nation-states. This is because increasing numbers of people live in what Said has called 'a generalized condition of homelessness' (Said, 1979, p.18) – what Salman Rushdie refers to as 'imaginary homelands' (Rushdie 1990): a world in which identities are less closely married to residence in specific 'national' territories. Refugees, migrants and travellers are the most obvious populations of these homeless people, but there is also a more general issue at stake here, as the anthropologists Gupta and Ferguson have noted:

> In a world of diaspora, transnational cultural flows and mass movements of people, old-fashioned attempts to map the globe as a set of culture regions or homelands are bewildered by a dazzling array of post-colonial simulacra, doublings and redoublings, as India and Pakistan apparently reappear in post-colonial simulations in London, pre-revolution Iran rises from the ashes in Los Angeles, and a thousand similar cultural dramas are played out in urban and rural settings across the globe.
>
> (Gupta and Ferguson 1992, 7)

Seen in these terms, globalization is encouraging a simultaneous fragmentation of places and identities. Not surprisingly, some commentators are now wont to talk of *glocalization* to capture the Janus-faced nature of our modern world (see, for example, Lash and Urry 1994).

It might be contended, however, that the territorial 'nation-state' still remains strong, and that it is most obviously so in the realms of defence, warfare and patriotism. In many respects this is indeed the case, but even here the powers of particular countries are often changing and are not all that they seem.

According to Mark Zacher:

> While the traditional Westphalian system [of sovereign territorial states] did have an element of order that was based on states' mutual

recognition of each other's sovereignty, there was never a strong feeling in Europe that wars were or should be illegitimate. War for purposes of territorial aggrandizement or a desire to change foreign governments was quite common – 'a perfectly legitimate instrument of national policy'.

(Zacher 1992, 67, quoting Morgenthau 1989, 255)

In many parts of the ex-Second and Third Worlds this continues to be a widely supported philosophy, as witness the recent excursions of Iran and Iraq into one another's territories, and as witness the continuing Chinese occupation of Tibet. In the heartlands of modern capitalism, however, this is not the case, at least in regard to one another. The geography of modern warfare has changed significantly since 1945 (O'Loughlin and van der Wusten 1993).

There are two main reasons for this 'rewriting of the map of the world', both of which have regard to what Zacher calls the cost–benefit ratio of warfare and the nature of modern weapons systems. Within the so-called First World the incentive of any one country to attack any other has diminished since 1945 because of the evident dangers associated with nuclear weapons and mutually assured destruction, and because 'of the fragility of modern technological civilization, the higher priority assigned to economic welfare, the higher costs of military occupation in the age of the nation-state, and liberal democratic inhibitions' (Zacher 1992, 68). None of this means that wars within Western Europe, Japan and North America are now impossible, but they are unlikely to occur in the foreseeable future. Zacher also notes that this unwillingness to risk national self-immolation sits uneasily with theories of international relations that put a premium upon the *relative distribution of power* as a disposition to given state actions (see Chapter 6). Kenneth Waltz is especially associated with this realist tradition of IR, but in 1990 Waltz was obliged to recognize that:

because so much explosive power comes in such small packages, the invulnerability of a sufficient number of warheads is easy to achieve and the delivery of fairly large numbers of warheads impossible to thwart ... [the] absolute quality of nuclear weapons sharply sets a nuclear world off from a conventional one.

(Waltz 1990, 732; quoted in Zacher 1992, 69)

War in the traditional sense, and thus IR in some traditional versions, is becoming obsolete (see also Kaysen 1990).

In other parts of the world the prospect of 'no war' is a long way off. Indeed, the number of wars fought in the so-called Third World escalated consistently in the 1950s, 1960s and 1970s, in part in response to the ambitions and capabilities of the superpowers that used their lands for proxy wars, and more especially because of the massive investments made by the governments of many such countries in modern weapons systems. Waltz suggests that 'waging war has more and more become the privilege of poor and weak states' (Waltz 1990, 734) and, allowing for irony, this is partly true. It would be more accurate to suggest, however, that wars are now mainly sited in poor countries and that modern weapons systems are invested in above all by poor countries with coercive governments, often for use against their own poorer populations (Chomsky 1992). In some quarters the hope is that a possible diffusion of nuclear weapons to some ex-socialist developing countries will discourage warfare in these regions. But this is not the only scenario with which military theorists are working. A majority of developing countries will long remain without nuclear weapons and those countries which do become nuclear powers cannot be relied upon to substitute diplomacy for the threat of warfare, even when that warfare is carried out with conventional weapons (van Creveld 1991).

One final set of points needs to be signalled here. Thus far we have been discussing the changing nature of warfare and its changing geographical incidence. Crudely summarized, we can say that the richer countries are unprepared to fight their wars 'at home', although many of them are quite willing to commit their troops abroad, either in unilateral actions or under the auspices of the United Nations. It is worth noting, however, that warfare has always threatened the environment, as well as the lives of men, women and children directly. In some quarters this division between peoples and places allows a new celebration of sophisticated weapons systems that supposedly target missile attacks with 'surgical precision' and which involve bombs – neutron bombs – that allegedly knock out only buildings and not surrounding civilian populations (Hewitt 1983). At other times, and more reasonably, concern has been voiced about the destructiveness of trench warfare, about the effects of napalm and agent orange, and about the dangers of a nuclear winter that would indeed ensure the destruction of the destroyer (Sagan and Turco 1990).

More recently still, this idea of a war against the environment has been taken further and in some quarters would now include the anti-environmental wars waged by global economic growth, modernization and development (Albert 1992; Bryant 1991; Dalby 1992). Again, the locus of concern tends to be the Second and Third Worlds, at least in the academies and corridors of power of the First World. In the former Soviet Union concern about the environment probably first surfaced in the West with the publication of dissident accounts about the eutrophication of Lake Baikal and other inland seas (Komorov 1980), but it was brought to the world's attention more dramatically with the disaster at the Chernobyl nuclear plant. The effects of this nuclear 'accident' were felt at least as far west as Ireland and the Celtic fringes of the UK, where large meat-animal herds were condemned as unsafe for human consumption on account of the high levels of radiation to which they had been exposed. Needless to say, the local effects of the Chernobyl disaster were much more tragic.

In the developing world attention has focused on two main and related issues: the destruction of tropical rainforests (Hecht and Cockburn 1990) and thus biodiversity, and the relationships between rapid economic and population growth and a process of global warming that is encouraged by the emission of CFCs and other 'greenhouse' gases (Henderson-Sellars 1991). The greenhouse effect, ironically, if not quite predictably, threatens to be doubly unfortunate for many poorer countries. On the one hand, many such countries are scapegoated for pushing global warming beyond acceptable limits, notwithstanding the fact that the USA alone emits as many CFCs as most South Asian and sub-Saharan African countries combined, and notwithstanding the fact that average per capita energy consumption levels in the USA in 1990 were more than twenty times those found in the world's low-income economies (World Bank 1992; see also Homer-Dixon 1991). On the other hand, the poorer countries are among those most likely to suffer the effects of global warming. Bangladesh is poorly placed to suffer a predicted rise in global sea-levels consequent upon global warming (Gleick 1989), and the drought-affected areas of Latin America, Africa and Asia will not welcome a local increase in ambient temperatures unless rainfall levels also increase. In contrast, parts of North America and Northern Eurasia probably would welcome a local rise in mean temperatures. New and more profitable crops could be grown in hitherto marginal farming regions.

The environment is undoubtedly a 'global problem' (Benedick, 1991; Deudney, 1990), and one that indicates the porosity of almost all nation-states, but, as ever, the costs and benefits of an emerging ecological threat do not fall equally on all regions and peoples (deLemos 1990). Environmental problems are changing the terrain of and for geopolitics, but they are not rendering it inconsequential or anything of the sort (Brown 1992).

'GLOCALIZATION' AND THE INTERNATIONALIZATION OF STATE ACTIVITIES

Glocalization, hegemonic instability and economic crises

The processes of glocalization and deterritorialization that we have just described were clearly associated with the emergence of a series of international economic and political 'crises' in the 1970s and 1980s. These crises expressed themselves pre-eminently in the arena of international monetary relations, where they took the form of a series of problems linked to inflation, debts and deficits.

It is probably unwise to search for a single cause for these crises, but the worldwide increase in inflation in the 1970s (see Figure 7.3) undoubtedly was promoted by US attempts to fund its war in Vietnam by printing more dollars

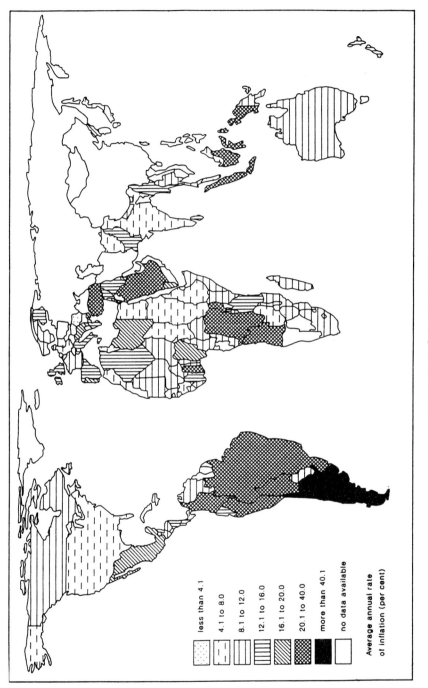

Figure 7.3 Inflation in the 1970s
Source: World Bank 1981

less than 4.1

4.1 to 8.0

8.1 to 12.0

12.1 to 16.0

16.1 to 20.0

20.1 to 40.0

more than 40.1

no data available

Average annual rate
of inflation (per cent)

(to finance a balance of payments deficit), and by the activities of certain money-centre banks engaged in recycling petrodollars to middle-income economies in Latin America and Asia. Such developments in turn were related to the emergence of new rivals to the economic might of the USA and to the willingness of the US, from August 1971, to pursue a politics of national exceptionalism that had little regard for the traditional stabilizing roles expected of a global hegemon.

By the late 1970s levels of inflation in the US and several other OECD economies had reached a point where the private capital markets were voting with their (increasingly large) feet against those currencies associated with 'profligate governments'. In September 1978 the markets took against the United States in a big way, and for perhaps the first time since the Second World War the USA had to reshape its economic policies to meet the demands of a group of economic actors lying outside its territorial borders. President Carter appointed Paul Volcker as the Chairman of the Federal Reserve Board in 1979 and money supply policies were tightened so sharply by Volcker that the prime (interest) rate in the US rose to 16.5 per cent in June 1982 (Greider 1987). The problem was that similar policies were also put into effect by Prime Minister Thatcher in the UK and by Chancellor Schmidt in West Germany (amongst others), with the result that a concerted attack upon global inflation by means of high real rates of interest led to the deepest economic recession since the 1930s.

This recession in turn created difficulties for a group of heavily indebted developing countries, many of which had taken out commercial bank loans in the mid-to-late 1970s to ride out the OPEC-inspired rise in oil prices *and* to meet Western demands for rapid economic growth in these regions of the (ex-)Third World. By 1982/3 a number of countries in Latin America were not in a position to meet some of the payments due on their debts and they duly defaulted. In Africa another group of countries defaulted on the official loans they had contracted to governments in the First World (mainly in Europe); just like the Latin American countries, they were unable to service their debts at a time of rising real interest rates and declining export markets. A debt-cum-banking-cum-export crisis was born only a few years after Walter Wriston had claimed that 'sovereign nations' could not go bankrupt (Wriston 1986).

The 1980s was a lost decade for development in many parts of Latin America and sub-Saharan Africa and the effects of the developing countries' debt crisis are still apparent in the early 1990s. From the point of view of the West's money-centre banks, however, the 'debt crisis' was handled quite adroitly by a policy mix involving export-led structural adjustment in the indebted countries, minor debt writedowns for the commercial banks after 1988, and a reflation of the US economy from about 1982/3. The reflation package in the US came to define that hybrid economic policy known as Reaganomics, which amounted to little more than a relaxation in domestic money supply policies run in tandem with a lavish fiscal policy. Guided by Art Laffer and others, the US administration came to believe that the total tax receipts of the federal government could be increased at a time of tax cuts by increased economic activity in the US economy at large (Stockman 1986). In practice, this optimism was not borne out and the federal government predictably began to run up a huge budget deficit as its spending outstripped its receipts from taxpayers and other sources. The budget deficit was associated with a burgeoning current account deficit from about the early mid-1980s, and the twin deficits were funded by a potentially unstable mixture of US IOUs to its trading partners and by massive US borrowing from foreign lenders (including Japan). A government that had come to power determined to make America 'great again' left office with a good deal of military hardware in place, but with a rapidly rising national debt that would be saddled on the backs of future generations of US citizens. Benjamin Friedman has concluded that Americans lived so well in the Reagan years mainly because each 'family of four' borrowed an average of $9,000 from foreign lenders in the period 1981–7 (Friedman 1987).

In still broader terms we can say that the US deficits (and the trading surpluses that emerged at the same time in Germany and Japan) had their origins, at least in part, in a set of economic and geopolitical obligations imposed upon the US government by the debt crisis in some Latin American countries. The blunt fact is that the US government could not afford to see several leading US commercial banks go to the wall following a concerted debt default by the big four Latin American debtors (Mexico, Brazil, Venezuela and Argentina), and so was forced, more or less, to reflate its own economy to hold out to these indebted countries the prospect of lower real interest rates and the possibility of export-led growth (Corbridge and Agnew 1991). The US clearly would have liked its 'partners' in Europe and East Asia to reflate their own economies at the same time, but this did not happen. Many of these countries were happy to be pulled along in the late 1980s by the locomotive effects of the US economy, but they were no longer willing to have their economic policies dictated by the particular needs of Washington and the US economy. The glocalization of economic activity, and the decline in some aspects of US hegemony, seemingly had created the conditions for a less regulated and more anarchic world economy in which the symptoms of disorder – from inflation through debts to deficits – were by 1990 more morbidly present than ever.

It is not easy to dispute the main contours of the story we have set out above and it is hardly part of our brief to do so. We clearly have been living through a period of disorder in international economic relations, as compared to the 1950s and 1960s, and this disorder has undoubtedly been associated with a decline in the powers of the US territorial economy in relation to other economic powers (including various institutions and the markets themselves).

191

These points clearly made, we do also believe that a new and very different 'order' is emerging in the contemporary geopolitical economy and that the map of this new world order is not only signposted by the crises of inflation, debts and deficits, or even by the food and political crises besetting parts of sub-Saharan Africa, or by a lost decade of development in many parts of Latin America. (Indeed, we want to argue that these 'morbid symptoms of disorder' can also – sadly – be read as signs of an emerging new 'order' for other actors and regions in a world economy increasingly regulated by market-makers, the circuits of capital, and internationalized states and quasi-global political institutions.) The 1970s and 1980s were also decades of unprecedented prosperity for the broadening upper-middle classes of many First World countries, and of extraordinarily rapid economic growth for those ex-Third World countries, classes and regions that had successfully negotiated a position for themselves in the global market-place. It is also worth emphasizing that these decades did not see the collapse of the capitalist world economy, or the capitalist banking and monetary systems, as several commentators had predicted at various points in this period. We need to recognize that a new world order is emerging which is surprisingly stable in its expanding core areas, and which has ridden out the very real threats presented by inflation, debts, balance of payments deficits and economic nationalism, but which is also 'orderly' in part because of a new capacity to write-off regions, countries and communities that are marginal to the development of this geopolitical economy.

In this latter regard, we find ourselves in partial agreement with David Harvey, who noted in 1982 that various crises in international monetary relations (and in the regulation of capitalism more generally) would

> unravel as rival states, possessed of different money systems, compete with each other over who is to bear the brunt of devaluation. The struggle to export inflation, unemployment, idle productive capacity, excess commodities, etc., becomes the pivot of national policy. The costs of crises are spread differentially according to the financial, economic, political and military power of rival states.
>
> (Harvey 1982, 329)

We also believe, with Harvey, that the real costs of the crises of the past twenty years have fallen most heavily on certain countries, regions and classes in the poorer regions of the 'Third World' (not least through the effects of structural adjustment programmes and market idolatry: Watts 1991), and on those classes and communities in the First and Second Worlds that have to live in the deadlands created by economic restructuring (Campbell 1993; Davis 1990). Where we differ with Harvey is in refusing the idea that these forms of crisis management must create further and *more damaging* crises for something as abstract as the capitalist mode of production. In some ways (but not all, see Chapter 8) our conclusions are more depressing. We believe that a new form of hegemony is now emerging which is decreasing the likelihood of economic and political collapse in so-called global capitalism, but which is creating new conditions for 'order' by ignoring, tidying away and/or disciplining a group of countries, regions and communities which are not party to a new regime of market-access economics or which threaten it in some way. This new hegemony is still policed in part by powerful countries like the US, Germany and Japan, but it is also held together by the markets (see next sub-section), by international institutions like the IMF, the World Bank, GATT and the UN and by a more diffuse network of overlapping powers and sovereignties that define a world of internationalizing state activities. It is to these institutions and networks that we now turn.

Internationalized states

A pure world of Westphalian nation-state actors has probably never existed. States have always had to share some of their powers with other states and Zacher notes that as early as 1909 there were '37 conventional inter-governmental organizations (IGOs) and 176 non-governmental organizations (NGOs)' (Zacher 1992, 65). As far back as the mid-nineteenth century, indeed, these organizations were sponsoring an average of two or three international congresses and conferences every year. Even so, the growth of internationalized state activities since 1945 (and even more so since 1970) has been quite staggering. Zacher's simple statistical analysis reveals that in 1986 there were 337 recorded IGOs and 4,649 NGOs, 'most of which prescribe a variety of regulatory arrangements for their members' (ibid.). Zacher also notes that:

> There has . . . been a marked increase in the number of international treaties in force in the postwar period. One study indicates that there were 6,351 bilateral treaties that entered into force in the decade 1946–1955; 10,456 treaties in the period 1956–65; and 14,061 treaties in the period 1966–75.
>
> (ibid.)

He continues:

> An interesting figure that indicates the growing importance of inter-national organizations is that in the first decade international organ-izations were parties to 623 treaties with states or other organizations; in the second decade they were parties to 1,051 treaties; and in the third decade they were parties to 2,303 treaties. It is also clear that the large majority of the treaties concerned international economic issues of various types.
>
> (ibid., drawing on Bowman and Harris 1984)

The internationalization of state activities provides a mirror to the internationalization of economic affairs in the post-war world system (Picciotto 1991). It is important to recognize, however, that the internationalization of states has not been even and that an apparent loss of sovereignty on the part of some states has been acceded to willingly and in recognition of changing realities, while in other cases, most obviously in poor and indebted countries, a loss of sovereignty has been imposed on weak or quasi-states (Jackson 1990) by (or in the name of) non-state actors like the IMF and the World Bank. The internationalization of the state system has plainly not yet been associated with an increased democratization of geo-economic relations, even though there is, we believe, some potential for empowerment along these lines (and some signs of democratization within poorer countries: Hawthorn 1993a).

In the core countries and regions of the new 'market-access' world order, the internationalization of state activities has proceeded largely by consent and negotiation. The process has probably proceeded furthest in the European Community (European Union), although there has been a recent backlash there against full monetary union. After 31 December 1992 a single market was created in the European Community and EC citizens have for some years now been able to find employment, and seek financial support for post-school education, throughout the Community with little or no regard for their nationality or place of birth. Member countries of the EC have for the most part been willing to cede some powers and revenues to the European Parliament and to the European Court, and the Community now intervenes regularly in the daily life of European citizens in matters as diverse as the law, health and safety policies, environmental policy and regional aid. It is significant, though, that the British government especially, with open and covert support from some other countries, has for some years been pushing for a European economic union while opposing a European cultural or social union, or super-state.

Much of the impetus for a greater Europe has come from business groups. Faced with an increasingly internationalized and deregulated world economy in the 1980s, international capital has mounted an aggressive campaign against those aspects of the EC that have threatened a free trade in goods and services in Europe. International capital has also led the campaign for a single European currency, but this campaign has stalled in the face of German reunification and the lack of convergence, as yet, between the national economies that still make up the greater EC economy.

The economic and geopolitical powers of the EC are clearly on the rise, notwithstanding some recent setbacks. The centrality of Germany in the EC also means that other member countries of the EC must have regard to German interest rates and money-supply policies when determining their 'own' macroeconomic policies (Lawson 1991). In other parts of the market-led world economy, too, countries are finding it increasingly hard to buck the global trend to deregulation, privatization and reduced government intervention in the economy. Even the most powerful countries now recognize that they can challenge the markets only in concert with several other leading economic powers. It is for this reason that the summit meetings (and backroom dealings) of the G7 and G3 countries have become so institutionalized of late, and so central to the regulation – and ordering – of a deterritorializing world economy.

In the 1980s the international debt crisis, and various associated balance of payments crises, were partly dealt with in such forums and by means of the Louvre and Plaza accords that were announced in the middle of the decade. These accords signalled the need both for concerted inter-state actions to confront a looming common crisis of international monetary disorder, and the willingness of most countries to act accordingly in such circumstances (Woodward 1992). By such means the much vaunted Prisoner's Dilemma effect in international relations was largely avoided; many leading countries recognized that the modern world economy could only be policed by means of a network of 'shared sovereignties' and with the support of like-minded international institutions like the World Bank and the IMF.

The likemindedness of the World Bank and the IMF became very apparent in the 1980s as these two institutions were empowered anew to deal with the threat of indebted non-development, or, more exactly, the threat of an international banking crisis. In the 1970s the Bretton Woods institutions had seen their resources dwindle in comparison to the resources of nation-states and the private capital markets, and the World Bank made use of those resources it did have available to support a project of development defined by Robert MacNamara and others as redistribution with growth and the provision of basic needs (Mosley, Harrigan and Toye 1992). At this time the idea of state intervention in support of 'development' and a social welfare function was not reviled. In the 1980s, however, a counter-revolution took shape in development theory and policy which had at least some of its roots in 'a change of mind in the North' where the US was 'uninterested now in pursuing its Cold War in the South' (Hawthorn 1993b, 18). The counter-revolution drew intellectual inspiration from the New Right and made much play of the alleged failures of dirigiste development hitherto (Lal 1983), and of the applicability of market principles in the countries of the South (Little 1982). Some proponents of the counter-revolution also saw in the East Asian NICs a proof of the virtues of market-led and outward-orientated economic development.

Politically, the counter-revolution was introduced to the Third World – let us put it no more strongly for the moment – by the IMF and the World Bank. These institutions have urged countries seeking their help to adjust to the global economic crises of the 1980s and to embark upon local processes of structural adjustment. As Hawthorn puts it, these countries are advised to 'sell off their unprofitable enterprises, cut their social spending, stop protecting their currencies, and open themselves fully to imports. They should, in short, restructure themselves' (Hawthorn 1993b, 18) and make themselves ready for life in the global market economy. According to the World Bank's *World Development Report* (1987, 35), 'there is no viable alternative' to this process of restructuring and structural adjustment. Things may have to get worse in the 'diseased' economies of many poorer countries before they get any better.

The medicine administered by the World Bank and the IMF has not been without its benefits – at least in regard to farm output pricing (Sender and Smith 1985) – and it was a medicine that was often administered with the support of many local élites in the poorer countries, as well as with the imprimatur of the US and its allies in the OECD. On not a few occasions it suited all parties for the Bretton Woods institutions (as they are still called, rather ironically) to draw the fire that would otherwise attach to particular governments and countries. Few would deny, however, that the medicine has been severe, and even committed neo-liberals must now be wondering whether the game has been worth the candle, or if more pragmatic policies might have worked less pyrrhic 'victories'. What is not in doubt is the geopolitical significance of this reinvention of the Bretton Woods institutions, and of the powers that can be imposed on different countries by the General Agreement on Tariffs and Trade. In many countries of the 'developing world' real control over interest rates and exchange rates, over fiscal policy and monetary policy, and even over the environment and local product mixes, has been ceded to various international organizations and even to some NGOs. In such countries we should talk not so much about states as quasi-states (Jackson 1990). The link between territoriality and sovereignty is rapidly being undone.

Market-access and/or market idolatry?

The structural adjustment programmes that some would associate with the new pre-eminence of the Bretton Woods institutions are best regarded as part of the much larger project we have identified as transnational liberalism. Structural adjustment programmes are in place in many parts of the ex-Second World as well as in the indebted Third World, and a process of market-led economic restructuring has been popular for fifteen years now with governments in many of the world's richer countries. To understand why this is so, we first need to recognize that an ideology of market-access economics (and a hegemonic regime of transnational liberalism) has not just emerged in response to the realities of glocalization or the self-interest of various political and economic élites, although both of these factors are important. Transnational liberalism also commands *intellectual* support and it rests on a series of arguments that are not easily rebuffed.

In the 1930s Maynard Keynes reminded his readers that, 'the ideas of economists and political philosophers, both when they are right and when they are wrong, are more powerful than is commonly understood' (Keynes 1936, 383), and we are fully in agreement with Keynes' judgement. It is somewhat ironic, however, that this very insight was turned back upon 'Keynesianism' – both in theory and in practice – by an emerging school of market economists and libertarian philosophers in the 1970s. A diverse group of neo-liberal economists led by monetarists like Friedman and McKinnon and 'Austrians' like Hayek and Kirzner, and later by supply-siders and constitutional economists including Buchanan and McKenzie, found it convenient to make common cause against what they saw as the dirigiste fallacies of the post-1945 period and the misuse of Keynes' ideas by so-called Keynesians. Their celebration of the efficiency and transparency of market-led economic transactions also meshed easily with the libertarian revival led by Hayek and Nozick among others – a revival that equated democratization with the anonymity of the market (and minimal state interference) and which argued that 'Inequality [of outcomes] is notoriously the common condition – one might well say "the natural condition" – of mankind' (Minogue 1990, 100). For Minogue, as for Friedman, Hayek and Nozick, 'My disagreement with egalitarianism [as opposed to an equality of opportunity in the market-place] is thus both particular and general. I oppose not only the plan of equality, but any kind of plan at all' (ibid.).

It is hard to exaggerate how radical these views seemed at first in the 1970s. All of the thinkers we have referred to above were for many years marginalized by the political and economic establishments of the post-war era. In the Golden Age of Fordist Capitalism, planning was promoted as a new godhead in the West's rapidly secularizing societies. The Second World War was explained away as the result of the competitive economic nation-alisms of the 1930s and with reference to the anarchic and depressive tendencies that allegedly are built into the structure of market-led economies and economics. The work of Maynard Keynes and his followers was then read as an antidote to these destabilizing theories, policies and outcomes. Keynesianism was supposedly written into the political settlements of the post-war era in such a way that economic growth and increased welfare would be brought to the populations of the First World by governments acting to dampen down unemployment and the business cycle (Hall 1989). Economic modernization would in turn be supported in the non-socialist Third World by massive infusions of foreign aid and direct foreign invest-ment, and by intellectual and political support for Third World governments that invested heavily in private and public sector industrialization and their attendant infrastructures. The United Nations' Development Decade launched by President Kennedy in 1961 even suggested that the 'problem of development' could be eradicated in ten years. Latecomer societies could learn from the experience of the pioneer industrial countries and take-off rapidly into an era of self-sustaining growth and eventual mass consumption (as per the USA: see Rostow 1960). Such optimism was also mirrored in the Second World, where Khrushchev committed the Soviet Union to out-stripping the economic might of the USA by means of rational socialist planning and production, and in the heartlands of advanced industrial capitalism where Harold Macmillan, the Prime Minister of the United Kingdom in the late 1950s and early 1960s, told his countrymen and women that 'they had never had it so good'.

Macmillan was right, of course, but the extreme optimism of this period was soon to come under fire. The 1960s was a pivotal decade in the history of the post-war world system, as most commentators now recognize. In Czechoslovakia in 1968 the Soviet Union was again obliged to turn to force of arms to keep its subject nations in Eastern Europe in tow to a political and economic system to which many there objected. Meanwhile, America's good intentions in the 'Third World' were called into question again by the Bay of Pigs fiasco in Cuba and by the escalation of the war in Vietnam by Presidents Kennedy and Johnson. These two wars were critical in encouraging the growth of the *dependencia* school of political economy which linked the development of the rich countries to the exploitation and impoverishment of the world's poorer countries (Palma 1978; Hout 1993). Statistics from this period showed a widening absolute income gap between the world's rich and poor countries (Jalée 1969), and this further encouraged the consumption of *dependencia* ideas in the academy.

Finally, and if this was not bad enough for a capitalist world system that was supposedly moving beyond ideology and dissent, the economies of some richer countries began to slow down as creeping inflation 'infected' the Western world in the later 1960s and 1970s. In the 1970s this inflation increased further amid an expanded sea of xenomonies and petrodollars, and the economic hegemony of the West came under concerted challenge from a Third World demanding a New International Economic Order (Hoogvelt 1982). Economic nationalism (and nationalizations) probably reached a peak in the Third World in the mid-1970s, and in many advanced industrial countries real rates of profit declined in the wake of accelerating inflation, public spending and labour militancy. The optimism of the early and mid-1960s quickly turned to despair as the trade-off between low rates of unemployment and inflation predicted by the long-run Phillips curve gave way to an era of stagflation (Bruno and Sachs 1985).

It was against this background, which itself was associated with a decline in some aspects of US hegemony in the 1970s, that the seeds of a newly ascendant ideology of market-access economics and transnational liberalism were first sown. Hayek had long campaigned against the road to serfdom that he associated with a 'misreading and misuse' of some of Keynes' ideas by so-called Keynesians (Hayek 1944), but in the 1970s his ideas found political favour for almost the first time with the Thatcher–Joseph wing of a radical-izing Conservative Party in the United Kingdom. Slowly at first, but soon with more zeal, the idea gained ground that increased government spending was crowding out private sector investment and was discouraging the all-important quality of entrepreneurship (Gamble 1988). Whole communities of citizens were also being drawn into a dependent relationship with the state that was at once stultifying and disempowering. If real freedom was to be fostered, this argument ran, real capitalism would have to be re-invented and the role of the state in capitalist societies reduced to its proper functions of providing for defence, law and order and establishing property rights. This was the message of Milton Friedman's *Capitalism and Freedom*, first published in 1962 and later reissued with a new Preface in 1982, and of Robert Nozick's more subtle analysis of *Anarchy, State, and Utopia* (first published in 1974).

This general message was in turn reinforced by a corpus of more technical work in economics, which seemed to lend support to three important conclusions: (1) that inflation was always and everywhere a monetary phenomenon that was caused by governments printing too much money (in part to buy votes and/or to buy off political opposition from sundry militant groups); (2) that rising inflation would create so much 'noise' in supposedly market economies that long-term economic decisions would inevitably be distorted by a lack of 'true information' about the state of labour, money and other markets; and (3) that, while governments should not seek to keep local unemployment levels below the non-accelerating inflation rate of unemployment (NAIRU), they could expect unemployment to fall quickly in the wake of stern pro-market and sound money policies as workers rationally adjusted their expectations about inflation and the wider economy. These three sets of propositions were associated especially with the theoretical and empirical work of Friedman and Anne Schwartz (Friedman and Schwartz 1963; 1982) [(1) and (2)], and with the work of Michael Barro and Robert Lucas (Barro and Grossmann 1976; Lucas 1978), two labour economists working within the 'new' framework of rational expectations economics [(3)].

In somewhat broader terms, we can say that the main aim of the neo-liberal 'counter-revolution' in economic theory and policy (Toye 1993) was to link a period of global economic sclerosis in the 1970s with a mixture of socialist, dirigiste and Keynesian policies, allied to a politics bound up with social democratic ideas of equality and citizenship. A second aim was to signal a way out of this 'crisis' by means of a new holy trinity of ideas and policies: capitalism, freedom and the market. It is also important to note that the market was fêted not just for its capacity to promote economic efficiency in a state of general equilibrium; the market was also acclaimed by Hayek and fellow Austrian economists as a 'spontaneous order' that offered the best available means of signalling economic information that is by definition scarce, uncertain and decentralized in origin (Hayek 1960). Hayek also argued that open and competitive markets are the best means by which to preserve the liberty of the individual. Competitive markets perform this function by enshrining the sovereignty of the consumer and by removing from states and other monopolies the power of arbitrary resource use. Markets then, to adapt a favourite aphorism of Hayek's, are the result of human action, but not of human design. Markets are not about end-states or intentions at all, but are about a continuing state of change wherein new 'possibilities and preferences [are discovered] that no one had realized hitherto' (Garrison and Kirzner 1989, 123). In short, markets offer a guarantee against the corruptions of government (and Leviathan), and they embody the most reasonable way of dealing with, and making sense of, a world based around fluidity, flows, change and movement. States, in this discourse, are about stasis, sediment-ation and distortions; markets offer an antidote to such self-willed sclerosis and entrenched hegemonies.

This is not the place to go further into the intricacies of the intellectual bases of neo-liberalism. Nevertheless, we do believe it is important to insist that the new hegemony of transnational liberalism that is taking shape in the global geopolitical economy is not only the product of material forces or interested élites. Politicians and scholars who are not attracted to neo-liberalism will only deceive themselves if they believe that neo-liberalism is not underpinned by a strong set of intellectual arguments, or if they believe that many lay people do not implicitly register the force of these arguments. The success of Ronald Reagan, the popularity of council house sales in the UK, and the break-up of the Soviet Union (and the seeming demise of a socialist economic alternative to the market), all owed something to the apparent vitality of the market economy and market economics. The developing crisis of the welfare state in many richer countries is also being addressed most coherently – if not attractively in our judgement – by the ideologues of transnational liberalism. Similarly, in the ex-Third World, the heyday of nationalism and nationalization is long past and there is a growing willingness in many countries and regions to rethink state–market relations with respect to economic growth and development. The early and rather crude polemics against dirigiste development penned by Bauer and Lal (Bauer 1972, 1981, 1991; Lal 1983) have been superceded by the more sophisticated monographs and reports of Anne Krueger and the World Bank, amongst others (Krueger 1974, 1985, 1993; World Bank 1981, 1985). This body of work provides important empirical material on the tendency towards rent-seeking and urban bias in many of the countries beset by planned development (Bates 1981; 1988). It has also contrasted the pessimistic assumptions and outcomes of continued import-substitution industrialization (as in India in the early 1980s) with the success of outward-oriented development strategies which have acknowledged that markets can work (however imperfectly) in most developing countries. For such authors, a shift to market-led development will produce not just economic growth and increased opportunities in the medium-to-long term, but also a sense that local people can take control over their own lives (Colclough 1991). Instead of highlighting the dull compulsions of the market to which many critics of 'market idolatry' are keen to draw attention, proponents of neo-liberalism see in *the market* a site of empowerment that provides for equality of opportunity and that maximizes a local comparative advantage. In those countries where the central state is little more than a kleptocracy, as in Mobutu's Zaire, it is not hard to see why such arguments might hold an appeal (Bayart 1993; Hawthorn 1991).

None of the above means that an ideology of transnational liberalism has not been promoted by self-interested economic and political groups, or that a coherent challenge cannot be mounted against the worst forms of modern market idolatry. (We return to this in Chapter 8.) Indeed, there is no need to oppose self-interest to a general climate of ideas in this way. Modern conceptions of power and hegemony make the point that the interests of élites are most efficiently served when they refuse an obvious coercion, but rather appear to intersect with a generally accepted (or difficult to refute) climate of ideas that holds out the prospect of riches for many other communities and regions besides. If this section of Chapter 7 seems to emphasize the role of ideas rather than events and social groups in the production of a new hegemony, it is because the main tenets of neo-liberalism are not always spelled out in the IR literature and because we have dealt with the other two points repeatedly throughout this book. In the present context, we can deal rather more quickly with the role of dominant élites and with the practical implications that globalization holds for an ideology that links economic development to a disposition to gain access to *the market*.

The main point we want to make with regard to élite support for transnational liberalism is that any such élite must now itself be seen as a transnational entity (or set of interlocking interest groups). A thorough history of the geopolitics of transnational liberalism remains to be written, but it is likely that it would start in countries like Canada, where tight controls on the money supply were imposed in the mid-1970s, and in the UK where the Conservative Party in the late 1970s moved steadily away from the social democratic consensus that had been fashioned in post-war Britain. The governments of Margaret Thatcher (1979–90) were more pragmatic than some commentators have suggested, but there can be no denying that the rhetoric of Thatcherism in the 1980s did bring together many of the ideas of market economics and libertarianism that we outlined previously, and that Thatcherism in practice was associated with processes of denationalization, the deregulation of financial markets, and a general restructuring of labour markets.

There can be no doubt, too, that Thatcherite policies, much like Reaganite policies in the USA, worked to the particular advantage of well-to-do households (upon whom taxes were reduced dramatically), even though a majority of US and UK households improved their living standards some-what throughout the 1980s. Inequalities between rich and poor households widened in both countries in the 1980s, but the Thatcher and Reagan governments were careful to ensure that it was the poorest two deciles of the population in each country that lost out in absolute as well as in relative financial terms. A so-called underclass was allowed to emerge in the UK and the USA, and this 'class' was encouraged (by the poll tax link to voter registration in the UK, and by endemic voter non-registration in the USA) to cut itself off from the political process in both countries. When the marginalized populations did begin to make their voices felt it was often in gestures of despair and defiance (riots, anti-police actions, rising crime and drug use), and this was met in both the USA and the UK by government 'law and order' policies that criminalized these subject populations and that preached a new doctrine of 'rights and *responsibilities*'. If these populations had a problem (or problems), it was largely of their own making. The way out of crime and poverty lay in the market-place and by the individual taking responsibility for his or her own actions. There is reason to believe that large parts of the enfranchised population in the USA and the UK were happy to subscribe to this libertarian creed.

In France and West Germany post-1982 a not dissimilar resort to the language of market-access has been promoted by the dominant proprietary élites, albeit with more regard for West European vocabularies and traditions of welfarism. In Australia and New Zealand neo-liberalism has bitten even harder than in the UK and USA (Daly and Logan 1989), whilst in Japan the dangers of welfarism were highlighted by local élites mainly in regard to pensions payments and a growing population of retired men and women. (Government spending as a percentage of GDP in Japan has never reached West European levels, but is rather closer to the level set in the USA. Workers and families in the formal sector of 'big-business' Japan have traditionally been looked after very well, but generous welfare payments have not extended far into the larger population of 'informal sector' workers in the country.)

Perhaps more significant has been the growth of support for market-access ideas and policies in parts of the ex-Third World, and more recently in the former Soviet Union and in central and Eastern Europe. Writing of the former regions and countries, Biersteker suggests that:

> The bureaucratic, authoritarian regimes of the past were increasingly unable to deliver the promise of development (on which their legitimacy was principally based). Therefore, new interests have increasingly come forward to challenge the prior bases of statist, redistributive, inward-oriented, authoritarian regimes. There has, therefore, been an interest-driven demand for a major change in economic policy [in the ex-Third World], derived at least in part from a populist rejection of statism, and partially from the emergence of new economic elites dissatisfied with

excessive state interventions in the economy. . . . Local business elites have a great deal to gain from economic liberalization and are today more likely to demand a reduction of state intervention and more room for their own expansion.

(Biersteker 1992, 114–15)

Biersteker is undoubtedly right to write in these terms both about the ex-Third World and, by implication, about the ex-Second World. To this characterization, however, we want to add three more general points and observations. First, it is important to emphasize (as Bierstaker does) that many governments in the ex-Third World are trying to roll back state intervention in the economy because ineffective state interventions there have worked to their *political disadvantage*. In a country like India, import-substitution industrialization (ISI) will only work if the country has a strong executive state (as it did in the 1950s) that can disregard particular interest groups and can secure continued public monies for investment in develop-ment projects that often take decades to mature and show a profit. If and when the strong state is co-opted by powerful elements in civil society (as it was in India in the 1960s by upper level bureaucrats, richer farmers and the (monopoly) industrial bourgeoisie: Bardhan 1984), then ISI will often show a tendency to corruption, sclerosis and involution that brings the state as employer into conflict with dissatisfied consumers (of overpriced fertilizers for example) and disgruntled and underpaid employees. Unsurprisingly, the government of India today, like that of Prime Minister Thatcher in the UK in the 1980s, is keen to take large parts of the state out of the economy. The government can then create the appearance at least of labour disputes being a matter for the private sector (World Bank 1989).

Second, the interest-driven account of neo-liberalism put forward by Bierstaker clearly meshes with other accounts of the rise of a transnational business class. Writing of Latin America and parts of Africa, David Becker and Richard Sklar have for some years maintained that in our post-imperialist world many Third World élites speak a language of populist economic nationalism at home – down with the wicked IMF! – even as they transfer their own monies abroad for fear that populist policies will devalue their domestic assets (Becker and Sklar 1987). For such internationalized groups the debt crisis of the 1980s was probably more a blessing than a curse (MacEwan 1986). The debt crisis allowed some of these élites to blame the poor macroeconomic performance of a country on outside interests (the US and its banks, and the IMF again), while maintaining the values of their own monies in offshore bank accounts and welcoming the day when IMF-inspired structural adjustment programmes would create a domestic economic climate that could allow them to bring their fortunes home (Dornbusch and Edwards 1991).

Third, and finally, we think it is important to point out that the expanded scale of today's transnational bourgeoisie, and its erstwhile imitators, is largely the result of a prior era of state-led developments that built up industries and an infrastructure in the Third World, and that has encouraged some to believe that poverty is a thing of the past in the so-called First World. To put it bluntly – and in the form of just one example – the sale of council houses to working-class Tories depended upon a prior era when public sector housing was built by Labour and Tory governments pursuing precisely that brand of welfarism that is now reviled. Such are the ironies – the selective memories – upon which the new order of transnational liberalism depends.

A final set of supports for transnational liberalism is bound up with our concerns in the first half of this chapter: namely, the internationalization of capital and the technologies of globalization. We will argue in Chapter 8 that there are alternatives to a vision of market-access economics that too easily skirts the differences between the open and symmetrical markets celebrated in economic theory, and the inaccessible and asymmetrical markets that so often define real markets in the modern world economy (Altvater 1993). Markets can be transparent and empowering, but they rarely are at present (Miller 1989). Putting any such alternative vision into practice, however, will not be easy. Alternative policies do first depend upon alternative ideas and vocabularies – as with Keynesianism in the 1930s and feminism more recently – but they also have to take shape within and against a prevailing geo-economic climate. If our general argument is right, a challenge to market idolatry will then have to challenge, or to redefine, those processes of internationalization and globalization that have paved the way for a new era of transnational liberalism.

Again, we have to acknowledge that this will not be easy and it may not be possible to reclaim powers from *the markets* on a country-by-country basis. The experience of the past fifteen years shows that even powerful countries, like France in the 1980s, find it difficult to set national economic and political agendas that run contrary to the interests of private international capital (*the markets*) or the new ideologies of 'free'-market economics. Many poorer, or smaller, or weaker countries find it harder still. The fact is that exchange rates and interest rates, together with fiscal and monetary policies more broadly, are now being determined for some countries by *the markets*, and by the huge funds that have accumulated in the hands of the key market-makers like insurance companies, manufacturing TNCs, banks and pension funds. The main responsibility of these institutions is to their shareholders and investors. In a sense, indeed, their main responsibility is to the ethic of market capitalism itself, and to the priority that newly mobile capitals have established over relatively stationary nation-states (Fagan 1990).

None of this means that the new system of international relations that is emerging is untenably unstable, even if it is less obviously based around countries and nations. Market-makers must also have regard to the stability of the market-system and there is a good deal of evidence emerging to suggest that a cataclysmic collapse of the global market-place is far from likely, despite some early expectations to the contrary on the Left. To the extent that markets are themselves information processing systems, and even 'learning experiences', it is unlikely that the great traumas of the past (such as the Great Depression of 1929–33) will repeat themselves in similar forms. A world of debts and deficits hardly suggests that the market economy is stable in one sense of that word, but it is not implausible to suggest that new forms of 'stability' are emerging in a market economy that thrives on – and indeed is defined by – change and instability (Thrift and Leyshon 1988). The new stability, such as it is, might take the form of a further inter-nationalization of state activities to part-regulate the global market-place (as through the G3 and G7 Summits), and by more piecemeal attempts to diffuse risks within particular markets by means of securitization, futures contracts, anti-cyclical computerized share-dealing programmes and so on. In a new world order of internationalized states and private capitals, we have to begin to rethink our traditional accounts of hegemony, order and stability. An expanding core of market-accessible regions might well survive in spite of – even because of – the instabilities that scar the 'peripheries' of this new hegemonic network.

CONCLUSION

It is as well at this point to take stock of some of the arguments set out in this chapter. Our chief concern has been to outline the contours of a new hegemonic regime of *transnational liberalism*. The conditions of existence of this new hegemony lie in two main areas: in the facts of glocalization, and in the attendant ideologies of neo-liberalism and market-access economics. During the Cold War Geopolitical Order hegemonies were broadly imposed by two leading 'nation-states', the USA and the USSR, and the principal concerns of the international relations community were (1) to provide an account of the international order/stability that was made possible by this system of hegemony, and (2) to examine the military balance of power/terror that underpinned the Cold War Order. In our judgement, such a state-centric perspective on international relations will no longer suffice. The new hege-mony of transnational liberalism is both polycentric and expansionist, and possibly unstable (in some respects) as a result. It is polycentric because power in the modern geopolitical economy is no longer (if it ever was) monopolized by nation-states. Economic, cultural and geopolitical power is now embedded in a network of dominant but internally divided countries (including the USA, Germany and Japan), regional groupings like the European Community (European Union), city regions in the so-called Second and Third Worlds, international institutions including the World Bank, the IMF, GATT and the United Nations, and the main circuits and institutions of international production and financial capital. What binds these diverse regions and actors together is a shared commitment to an

A

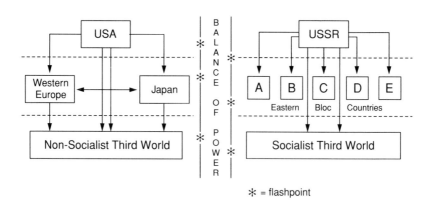

B

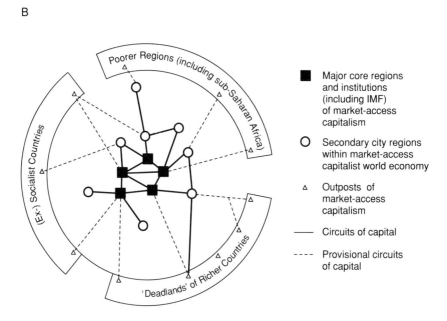

Figure 7.4 The reshaping of hegemony after the Cold War
A Hegemony in the Cold War Geopolitical Order
B Hegemony in the post-Cold War Geopolitical Order

ideology of market economics and a growing recognition that territoriality alone is a not a secure basis for economic or geopolitical power.

Figure 7.4 offers a stylized view of how this new hegemony is taking shape, and how it differs from the hegemony of the Cold War Geopolitical Order. Such a diagram is designed to emphasize the differences between the hegemonies of then and now, and the reader should bear this in mind when consulting it. It is certainly not our intention to suggest that elements of a dispersed hegemony were not present in the 1970s, or that a concern for strategic issues and balances of power/terror is somehow unimportant today; this is far from being the case. Our main purpose in Figure 7.4 is to underline the need to rethink traditional IR accounts of order, hegemony and stability in the wake of the *glocalization* of economic activities, and the inter-nationalization of some state activities, that we documented in the Chapter proper. A coherent account of international relations in the 1990s has to begin not only with a reprise on the end of the Cold War, but with an extended discussion of sovereignty and order/disorder in the wake of a continuing decentring of economic activity and a related deterritorialization of the geopolitical economy. To put it one last way, the emergence of new *spatial practices* demands that we rethink our *representations of space* **and** our prognoses concerning possible *representational spaces*.

This applies particularly to concepts of order and hegemonic stability. In this chapter we have tried to show how a new kind of order/disorder is being created in the modern world system. It is an 'order' that threatens to mimic the anarchic intellectual logic of the market upon which it depends and which it helps to secure. The new hegemony threatens to be almost as diffuse, with economic and geopolitical powers being shared unevenly between an over-lapping network of discrete geographical power centres and space-spanning 'sovereign' bodies. We have also indicated that this new hegemonic order can tolerate a worrying disorder on its borders and that, indeed, a willingness to write-off or police those regions and communities ('wild zones') not subscrib-ing to or incorporated into the market-access regime is built into this new hegemony. Chapter 8 takes up this question at greater length. It also reflects on how a (anti-) geopolitics based around *real reciprocity* might be advanced as a countervailing thrust to the new hegemony of the market and the powers of global capital.

Part III

The Elements of a New Geopolitical Discourse

8

MASTERING SPACE OR EMPOWERING COMMUNITIES?

Structures of hegemony have always emerged out of a geopolitical struggle by dominant states and their ruling social strata to master space – to control territories and/or the interactional flows through which modern terrestrial spaces are produced. The intellectual traditions of neo-realism and neo-liberalism take this as given. The first does so through its invocation of the pursuit of primacy, while the second does so through its claim that markets always reward the most efficient and punish the slothful and profligate. In practice, the end result has been a *coercive hierarchy* of 'greater' and 'lesser' states and regions, or of 'more' and 'less' 'developed' economies. In neither case is there the prospect of *real reciprocity* (to be distinguished from that usage which posits a congruence between economic benefits and military burdens, see, for example, Denoon 1993) and majority empowerment: *a counterhegemonic order based on a broad diffusion of power among a large number of actors committed to a set of universal principles of reciprocal interaction without entrenched domination.* Open and competitive markets might promise a world in which space is lived rather than mastered, but today's market idolatry belies this hope.

Of course, from the conventional points of view, a world in which space is lived is Utopian; a dream-like future that is impossible to realize in practice. In this chapter we want to suggest that this need not be the case. The new world order that is taking shape after the end of the Cold War is by no means a stable one or one bound to deliver a global community of empowered citizens and disempowered hegemons. Several years into the Clinton administration it is clear that the USA is beginning to retreat from the multilateralism that Bush essayed at the time of the Gulf War and that Clinton continued in Somalia. Having stoked up a number of 'civil wars' in Africa and elsewhere during the Cold War (through support for proxies like Unita in Angola), the USA is now ready to retreat from its obligations and responsibilities in parts of the world that no longer matter to it *economically* or *geo-strategically*. It is also clear that Western demands for democratization in the ex-colonial world are largely rhetorical and are easily countermanded by the commercial imperatives of trade and investment links with countries like China and the

East Asian NICs. The commands of money are now so loud – money being the natural language of market-access economics – that even Mobutu, in Zaire, is prepared to set up 'opposition parties' in the hope that such a travesty of democratization will find its financial reward (Bayart 1993). Yet even recognizing this (and more), we do still believe that the relative decline of US hegemony, coupled with the recasting of hegemony itself through the circuits of capital, present conditions that are more propitious than before for establishing a 'post-hegemonic' order that is opposed to the domination of a few key states or the commands of a singular (and singularly asymmetrical) *market*.

To lend some focus to our discussion of a possible politics of post-hegemonic empowerment (or *real reciprocity*) we will return to the four principles of the new geopolitics that we described in Chapter 1. This allows us to look afresh at the relationships between globalization and local attempts to defend diversity and secure a meaningful equality of access to world markets. It also encourages us to point up the possibility for political actions that challenge the sense that history is simply about repetition and the succession of one great power by another. Taken together, these two sets of observations, each of which is based on the breakdown of established *representations of space*, allows us to lend our voices to certain new *representational spaces*. Simply stated, these new representational spaces are informed by the diverse discourses of market socialism, internationalism–multilateralism, and a series of more local oppositional imaginations that seek to empower groups of subaltern men and women. Before addressing each of these issues, however, a more abstract consideration of the term 'mastering space' is in order, given its centrality to the logic of all uncontested global hierarchies.

MASTERING SPACE

Hegel (1931 edn), in his famous discussion of 'lordship and bondage', insisted that a single consciousness could know itself only through another, even in a situation of radically unequal power. The lord is lord only through the relationship with a bond servant. In the relationship, however, the other is sublated or annulled. The self of the lord derives from the conquest and *negation* of the bond servant. At the same time, only recognition of the selfhood of the other allows for its negation. Ironically, therefore, lordship or mastery implicitly recognizes the separate selfhood of the dominated. They are nominally equal selves locked into an unequal and enduring hierarchy.

Metaphorically, Hegel's model of lordship and bondage is the frame of reference in which conventional understandings of international political economy operate. But it is not the essential equality of the selves that has received most attention. Rather, it has been the assumed naturalness of an enduring hierarchy. In a world of competitive states, or a competitive world economy a 'singular' appropriation of power and wealth is taken as a natural feature of a world in which individual states/persons/firms objectify everything that they see as other, and where they view other states/persons/firms as little more than obstacles in their path.

There is a point of connection here between this conception of the mastering of others-elsewhere implicit in the logics of previous geopolitical discourse and practice and a feminist reading of Hegel's dialectic. From this point of view, to accept the unchanging polarity of man and woman in the lordship–bondage relation is to succumb to a structure of male domination. Hegemony can be seen as resting in both cases as much upon *negation* as upon positive recognition for its dynamic force. And, 'The more negation can be inscribed in silence, the more binding it will be, for the explicit denigration of women [or others] constitutes an act of naming that permits differing views about the strength of the tie between name and thing, between sign and referent' (Fox-Genovese 1991, 237). Awareness of this negation, therefore, is the first step out of the polarity upon which coercive hierarchy relies for its material and discursive power.

Polanyi (1957) historicized Hegel's dialectical model in his interpretation of European economic history in the nineteenth century. In the first phase of a 'double movement' came the self-regulating competitive market; what Polanyi viewed as a Utopian vision backed by the state. In a second phase came society's unplanned reaction of self-preservation against the depradations of market-oriented behaviour. In the guise of political regulation and the welfare state, society set about 'taming and civilizing the market' (Cox 1992). Hegel's model is potentially subversive, therefore, when put into historical motion. A singular hierachy becomes historically contingent. It is no longer necessary or inevitable for a singular state to play the part of hegemon, or for hierarchies to emerge that are rigid and exclusionary. Hierarchies of sorts will always exist in complex societies (based upon extended divisions of labour), but it does not follow from this that particular positions in a (changing) hierarchy must be occupied for long periods by the same economic and political actors. Hierarchies can be more or less fluid or coercive.

In the late twentieth century a recurrence of the 'double movement' identified by Polanyi can be seen in a very different context. Now a dynamic of globalization is prising open national economies and in many countries the welfare state is being undermined by market manias. At the same time, the beginnings of resistance are visible, if yet without much coherence given the weakening of territorial states as vehicles for constraining the market. New social forces are emerging among environmentalists, women, indigenous peoples, peace activists, labour unions and churches, to name just a few of the groups that are concerned with the less beneficial impacts of globalization. The relatively powerless are still often fragmented by religion, gender, ethnicity and a resurgent nationalism – even where they are also empowered by these identities – but their presence all over the world indicates elements of a common subordination in the face of a developing global society. In this setting of *consciousness* of the global, unlike in previous times when a sense of spatial isolation and separation prevailed, the mastering of space by a few can be challenged by new types of geopolitical discourse that stress reciprocity, or the necessity of mutual recognition, respect and toleration of different social positions and practices. We want to conclude the book with a brief examination of these new discourses.

GLOBALIZATION AND REAL MULTILATERALISM

The globalization of the international political economy is based on a number of processes that we have documented already: the internationalization of production and finance, the new international division of labour, massive migration from poorer to richer countries and the internationalization of state activities. These changes have been accompanied by shifts in the nature of communication and informational technologies that collectively have been called the 'informational revolution'. The new technologies effectively tie together people in widely separated places. The pace and intensity of the movement of ideas, products and people have increased explosively since the 1960s with the development of jet passenger transportation, computer networking and electronic communications.

What is most important geopolitically about the displacement of boundaries by flows is the potential for undermining conventional notions of spatial hierarchy.

> Flows are decentering, despatializing [in conventional terms], and dematerializing forces, and they work alongside and against the geopolitical codes of spatial sovereignty. Within the flow, there are new universals and new particulars being created by the networks of accelerated transnational exchange as fresh identities, unities and values emerge from sharing access to the same symbols, markets, and commodities.
>
> (Luke 1993, 240)

One consequence of this is the creation of new transnational communities which blur the spatial definitions of 'them and us' and 'ally and enemy'. Groups of environmentalists, feminists and peace activists are perhaps the best examples of such communities.

A second consequence is perhaps even more revolutionary. The sense of linkage or connectivity among people who are widely scattered challenges the usual assumption that people necessarily care only about those who are spatially close to them. The distance-decay model of morality typically sees responsibility as arising out of proximity to the other as a 'face'. 'Proximity means responsibility, and responsibility is proximity' is how Bauman (1990, 184) summarizes this view. Charity begins at home and distance inhibits a sense of obligation to the needs and rights of *distant* strangers. But why should seeing in the flesh, so to speak, signify a more transparent grasp of intersubjective responsibility? Some of the greatest moral offences of the twentieth century, including the Holocaust of the Jews and the Chinese Famine of the early 1960s, have been visited on neighbours. The act of 'seeing' people close by is as socially constructed as 'listening to' or 'watching' those far away. In the contemporary world there is no reason to suppose that moral boundaries would coincide with the boundaries of our everyday community; not least because these latter boundaries are themselves not closed, but rather are defined in part by an increasing set of exchanges with distant strangers.

In terms of the new representational spaces that we alluded to earlier, we are pressing here for a new language of internationalism–multilateralism that is appropriate to a new era of globalization (or what Giddens calls 'radicalized modernity': Giddens 1990). In one sense, at least, there are no Others in the contemporary world. The lives of men, women and children in one part of the world are inextricably bound up with, *and defined by*, the lives of men, women and children in distant parts of the same world system. Moreover, the prosperity of what is still a minority population in global terms cannot reasonably be explained without reference to the history of imperialism or to the accidental geographies of birthplaces and life-cycle opportunities. People born into middle-class households in the USA will almost always fare better than equally talented individuals born into poverty in Bangladesh. It follows, we believe, that neo-liberal arguments against international aid and more radical reforms of the world trading system are misplaced. Precisely because, as Rawls has noted (Rawls 1971), no one deserves their talents (and therefore their rewards in a very general sense), it follows that richer people in rich and poor countries should be encouraged to transfer some of their wealth to poorer citizens who lack certain basic human needs and what Rawls, again, has called 'the means to self-respect' (see Williams 1993, 7). Neo-liberals might consider this to be a form of coercive taxation (Nozick 1974), but we cannot agree with this suggestion. Right-wing libertarians see almost all forms of taxation (other than to provide for their defence and law and order) as coercive! Presumably, everyone always gets exactly what they deserve so taxation is by definition a levy on talent and merit.

We further believe that Rawls' theory of justice can be extended to the international stage, precisely because a sense of a global community is now emerging (Beitz 1991). We also believe that this argument holds true notwithstanding certain well-known objections to recent practices and patterns of foreign aid delivery. Official development assistance (ODA) is clearly imperfect, but it can be improved upon, and recent evidence suggests that great improvements have been made in the 1980s in the kinds of foreign aid delivered (if not in the quantities of ODA made available to recipient countries: see Cassen *et al.* 1994; Riddell 1987; Mosley, Harrigan and Toye

1992). In our view, there is a strong case for aid to be multilateralized still further and for its transfer and local use to be supervised by Western and non-Western non-governmental organizations (NGOs) (such as Oxfam). There is also a strong case for a massive expansion of aid budgets to help rescue people from levels of absolute poverty not of their own making and which most people in the West would not be prepared to accept themselves. This expansion in ODA should be embarked upon in tandem with an attack upon the many inequities built into the global trading system. Indeed, the case for reform here is even stronger, given the hugely damaging consequences of present trading restrictions and given neo-liberalism's own commitment (ostensibly) to open trade and level playing fields. Finally, we would want to see any such resurgence in a spirit of economic internationalism complemented by a commitment to political multilateralism. In most Western countries the right to vote is not determined by, or in proportion to, a person's income or wealth. But this is what happens still in the World Bank and the IMF, and in a rather different way in the Security Council of the United Nations. This has to change. We recognize, of course, that in terms of traditional geopolitical thinking, it is a matter of he (or she) who pays the piper calls the tune. Our point is that an opportunity is arising for the languages of neo-liberalism and market-access economics to be turned back upon their sponsors in a more pluralistic global political economy. The facts of *glocalization*, when added to a political discourse that stresses real equalities of opportunity (and thus the meeting of basic human needs worldwide), sits quite comfortably with a new representational space that we will call *real multilateralism*.

GLOBALIZATION, IDENTITY AND REPRESENTATION

The trend from boundaries to flows that is associated with globalization is also bound up with a more general crisis of representation. In some respects this can work against the formation of wider forms of political consciousness. As David Harvey puts it:

> Spaces of very different worlds seem to collapse upon each other, much as the world's commodities are assembled in the supermarket and all manner of sub-cultures get juxtaposed in the contemporary city. Disruptive spatiality triumphs over the coherence of perspective and narrative in postmodern fiction. In exactly the same way that imported beers coexist with local brews, local employment collapses under the weight of foreign competition, and all the divergent spaces of the world are assembled nightly as a collage of images upon the television screen.
>
> (Harvey 1990, 301–2)

Harvey goes on to warn that this tendency to spatial collage can engender a localistic reaction to globalization in which place-bound identities contribute to the very fragmention that they come into existence to resist.

216

From this perspective, political thinking does not necessarily lead in the same direction as the transformations in international capitalism might (at one level) suggest. There are indications that time–space compression has encouraged localism and nationalism rather than internationalism. Harvey points to events in the 1980s: 'The resurgence of geopolitics and of faith in charismatic politics (Thatcher's Falklands War, Reagan's invasion of Grenada) fits only too well with a world that is increasingly nourished intellectually and politically by a vast flux of ephemeral images' (Harvey 1990, 306). Optimism about the links between globalization and internationalization–multilateralism must, therefore, be tempered by the knowledge that the effects of globalization are often assimilated into understandings that arose from the practices of another era. Proponents of a 'minimal (liberal) universalism' – such as we outlined above, after Rawls – also now have to face a barrage of criticism from some of the more extreme adherents to the many discourses of postmodernism and postcolonialism. *Reason* and *Universalism* are often put into capitals and dismissed as colonial projects by those who mistrust the so-called totalizing discourses of Modernity and the Enlightenment. We have to say that we have little sympathy with such extreme standpoints. The critical geopolitics that we wish to support is about deconstruction and reconstruction; it is not to be confused with an affected playfulness which is so opposed to what it calls Foundationalism that it sometimes veers towards a solipsism and narcissism of the intellectuals.

None of this means that we are fully in agreement with Harvey's remarks on the perils of fragmentation. Harvey's argument needs to be developed in two further directions. To begin with, it is important to note that globalization is not as secure as some of its neo-liberal proponents maintain. Harvey (1990) and Gordon (1988) have not been alone in emphasizing that globalization has been based in large part upon the international movement of financial assets to take advantage of the monetary turbulence that followed the breakdown of the Bretton Woods system. From this point of view the global economy is a limited and fragile edifice rather than the behemoth it often appears to be in many accounts of globalization. Indeed, its effects have been restricted largely to the already-developed world and certain specific locations with institutional advantages (authoritarian governments, infrastructures, energetic regional bourgeoises) elsewhere. Capital mobility is still not a truly global phenomenon, and capital's increased pace of movement can be read as a sign of weakness as well as strength. Gordon makes the point as follows:

As the rate of return on fixed investment in plant and equipment has declined and as global economic conditions have become increasingly volatile, firms and banks have moved towards paper investments. The new and increasingly efficient international banking system has helped to foster an accelerating circulation of liquid capital, bouncing from one moment of arbitrage to another. Far from stimulating productive investment, however, these financial flows are best understood as a symptom of the diminishing attractiveness and increasing uncertainty about prospects for fixed investment.

(Gordon 1988, 59)

This is an important point because it suggests that the agents of globalization, above all the multinational corporations, are not totally in the ascendancy. Rather, 'The global economy is up for grabs, not locked into some new and immutable order. The opportunity for enhanced popular power remains ripe' (ibid., 64). The very insecurity of globalization, therefore, lends it a transformative potential, as does its basis in new and potentially decentralizing information and communications technologies. There are signs already that textile workers in North and South America are beginning to collaborate, and consumer activists and environmentalists have well-established cross-border agreements and organizations (Edwards and Garonna 1991). The simple truth that works in their favour is that the world economy still needs workers and consumers.

A second departure from Harvey concerns the possibilities for empowerment through a so-called political localism. Harvey recognizes that globalization and fragmentation/localization are proceeding hand in hand – the *glocalization* we referred to in Chapter 7 – but he sometimes fails to acknowledge that globalization can also enable a positive local politics of resistance to the pains of development. It can also enable a welcome decentralization and democratization of politics that can reasonably be accounted *counter-hegemonic*.

Again, we do not want to be misunderstood here. We do accept, with Harvey, that traditional forms of labour struggles, say, can very often be disrupted by the hypermobility of capital and by the politics of place competition it inspires (Cox and Mair 1988). We also accept that some local struggles to defend places and identities against global capital can be articulated by political vocabularies that embrace racist, communal and other identities, and which oppose the pleasures as well as the pains of modernity. The recent partial capture of the *Chipko andolan* (tree-hugging movement) in North-West India by advocates of ecocentrism and regional closure points to certain tensions in this respect. Most local people (and local women especially) do want to protect the hills of Uttarakhand against commercial logging interests and scientific forestry, but a good many are now opposed to those political groupings that want to keep *all* new roads and dams out of the region. This point made, we would also point to the success of the *Chipko andolan* as an example of the new possibilities for political empowerment and democratic decentralization that exist in the wake of globalization. Chipko activists have made clever use of an emerging global network of en-

218

vironmental activists, and have presented their case forcefully to the World Bank and other funding institutions and to a global television audience. The protection of a fragile regional environment by Chipko activists has made use of various global institutions and flows to contest the centralizing powers of government within a given nation-state (India). Not dissimilar struggles are being waged the world over by women's movements, peace movements, human rights groups and so on. Globalization is not only a synonym for disempowerment: it creates certain conditions for democratization, de-centralization and empowerment as well as for centralization and standard-ization. Globalization opens as many doors as it shuts. Its seeming logics of enslavement can be turned against it without eschewing all of the possibilities that development brings with it.

GLOBALIZATION, CITIZENSHIP AND HISTORICAL CONSCIOUSNESS

The democratizing impulses present in globalization were perhaps first demonstrated on the wider international stage by the events of 1987–90. A principal feature of the end of the Cold War Geopolitical Order was the outburst of popular opposition to it on both sides of the ideological divide. For the first time in modern history a geopolitical order underwent its final collapse from blows delivered by masses of people rather than by political élites. It was the *capacity* of people to see the irrelevance of the Cold War slogans in a transformed world that finally undermined the Cold War order itself. Citizens were able to imagine and act on scenarios about the state of the world that did not simply reflect established norms. If the huge public gatherings in the squares of Eastern Europe and the Soviet Union were by far the most important manifestation of this public mobilization, knowledge of the Western peace marches and demonstrations in the 1980s served as an important stimulus to the possibility of civil disobedience as a strategy of political change.

Rosenau notes of the scenarios which inspired the dynamic events of 1987–90:

> The new aptitude for thinking and feeling in dynamic rather than static, cross-sectional terms may not lead to sound conclusions or greater sophistication about world affairs. . . . Yet, whatever their level of accuracy or logic, scenarios in the present era are likely to be more elaborate than ever before – that is, marked by longer causal chains and stronger feelings, by selective judgments, by connections between events, by feedback links, by the perception that action at one time and place will set limits to or create options for action at another time and place.
>
> (Rosenau 1990, 367)

Irrespective of levels of 'factual' information, which often remain limited, exposure to a wide range of news stories and images builds the capability to consider and assess alternative visions of the future to those implicit in the conventional wisdom. One effect of this is to intensify emotional investment by fixing blame or expressing approval for any particular situation. Perhaps exposure to opinion polls has encouraged this tendency! For example, issues such as apartheid in South Africa, 'ethnic cleansing' in Bosnia, and US–Japan trade have involved intensive public debates in Europe and North America about historical culpability that would have been unheard of in previous eras (Rosenau 1990, 368).

The employment of historical consciousness has encouraged a questioning of established authority. Not just 'have-nots' but also 'haves' call into question old habits of compliance. 'More is at stake . . . than the mere revolt of dissidents. At issue is a change in our way of thinking about the basic nature and function of *authority itself*' (Harris 1976, 1). The competence of established institutions and leaders to respond to demands has been reduced by the emergence of issues related to increased global interdependence. At the same time active citizens are less easily satisfied with stock responses. Legitimacy now presupposes performance. The Chernobyl disaster became testimony to the difficulty of taking government claims at face value.

> Increasingly, the legitimacy of political regimes (and hence their capacity to rule non-coercively) is judged externally and internally, less by the standards of divine mandate, revolutionary heritage, nationalism, or charismatic authority, and more by the performance criteria specified in internationally defined standards of human rights.
>
> (Claude and Davis 1988, 2)

The crisis of authority poses a particular dilemma for communitarian philosophers unhappy with the political liberal-universalism that is usually associated with claims to 'human rights'. Their yearning for ethical order usually leads back to ancient Athens and the re-establishment of local communities. MacIntyre, for example, has called for 'the construction of local forms of community within which civility and the intellectual and moral life can be sustained through the new dark ages which are already upon us' (MacIntyre 1981, 245). The extra-local is here read as dark and destructive, as opposed to dark **and** light, empowering **and** disempowering, as we would contend.

At first sight there is more than a hint here of cultural relativism, in which the only appropriate standards for judging conduct would derive their validity from local practices with no appeal to practices elsewhere. This would be erroneous, at least for MacIntyre if not for some other communitarians. In a later book MacIntyre (1989) announced his fervent opposition to cultural relativism. In this version the confrontation with justice and rationality is resolved by an authoritarian Church. Choice becomes axiomatic and not the result of deep reflection. The global crisis of authority is thus resolved by turning discursively to the medieval Christian Church and the Pope. MacIntyre may well have a point. The longing for a certain ethical order might indeed require the imposition of a papal or secular authority. But whether many will listen is another question. Consciousness of the ability to change history, to which Pope John Paul II contributed through his defence of the Solidarity movement in his native Poland, will not easily be squeezed into the canons of medieval scholasticism. Some collective commitments, just like some of the imagined communities of active citizens, are broadening out as others are narrowing down.

GLOBALIZATION, *THE MARKET* AND 'MARKET SOCIALISM'

If there is one feature of transnational liberalism that achieved sacral status in the 1980s it is the idea of *the market* as the arbiter of the just and the true. This has been the inspiration behind the privatization of state assets and the liberalization of national economies all over the world. The material basis for this was laid by globalization, but its intellectual currency also owes much to the fiasco of socialist politics over the past twenty years. The two main wings of this politics, social democracy and Leninism, became committed to ideas that proved to be historically contingent at best, rather than transcendentally necessary. When compared to collectivization, or the collective ownership of the means of production, the idea of the competitive market has a seemingly timeless appeal that makes it ideal for political polemics.

But all is not well in the world of *the market*. Mrs Thatcher several times declared that 'Keynes is dead' (Thatcher 1993), but it is not clear that the period of her rule – and this misguided rubbishing of a dead man's legacy – was associated with the greater glory and efficiency of free markets. Some Britons might have accepted the first Thatcher recession (in the early 1980s) as a necessary price for the British economic revival – nay miracle – that Thatcher promised was about to ensue, but any such masochism/optimism has since been dented by the second great recession of the late 1980s and early 1990s. The age of *market economics* has certainly brought untold wealth to some households and regions (and not just in the old First World), but it has also coincided with stagnant real wages for many in the West, with widening income gaps in countries including the USA, the UK, Australia, New Zealand and Germany, with the international debt crisis, with the US trade and budget crises, with massive balance of payments disequilibria, with soaring global unemployment, and with declining real incomes for a majority of the populations of sub-Saharan Africa and Latin America. Set against this, there has been rapid economic growth in China and East Asia, but it is not clear that such 'successes' are owed entirely to the incentive effects of free markets, or to the role of markets as coordinating mechanisms. Matters are a good deal more complicated than some market ideologues maintain, as we shall argue.

221

By the same token, it is not clear that all of the failures of the 1980s were market-inspired failures. We certainly do not agree with Paul Kennedy (1993) when he argues that the world is not well prepared for the twenty-first century mainly because of *natural* Malthusian-style population pressures. The clear-felling of forests in Amazonia has more to do with cattle ranching and property speculation than with population pressure, and most of the world's pollution is generated in regions where population growth rates are low. Market failures count for more than Kennedy allows. That said, at least some of the crises of the 1980s must be laid at the door of government failures and local wars. It simply will not do to blame market economics and the IMF for all of the travails of the 1980s in the under-developed countries of the South, any more than it will do to proclaim the market as a panacea for the world's ills.

The key distinction we need to draw is between the theory of market economics and the actual operations of real markets in real places. We believe that open and competitive markets do embody many qualities that commend markets as engines of economic growth and as possible means of empower-ment, democratization and the dispersal of political power. The main theoretical arguments in favour of such markets are as follows. First:

> [competitive markets] when they work well . . . are an excellent way of processing information, while simultaneously providing incentives to act upon it. . . . There is no need for long communications between retail or wholesale outlets and central planners giving daily updates on the shortages or surpluses that are appearing; no need for detailed central planning directives to productive enterprises telling them how much or how little to produce.
>
> (Estrin and Le Grand 1989, 3)

Second, competitive markets encourage innovative behaviour on the part of producers as they seek to maintain or improve their position *vis-à-vis* competitor firms. Third, competitive markets disperse economic power. 'People have a range of other people with whom they can deal; they are not at the mercy of an awkward manager or a recalcitrant clerk. If they do not like the service of a particular supplier, they can go to another providing a better quality service' (ibid.). This dispersion of economic power – and monopsonistic purchasing power particularly – might be of particular benefit to farmers in sub-Saharan Africa, where crops have often had to be sold at below market prices to government agencies protecting the interests of urban consumers and the urban political élite (World Bank 1981). Finally, a persuasive argument can be set out which suggests that competitive markets increase freedom and democracy. Competitive markets should promote freedom of choice in terms of goods and services, freedom of choice in the labour market (if the demand for labour exists) and freedom of expression. David Miller also argues that:

Markets permit industrial democracy [although they do] not necessitate it. The extent to which it is possible depends on the structure of enterprises, with workers' co-operatives being the most democratic form. Such an enterprise structure may not be optimal for all industries at all times, in which case we face a trade-off between economic efficiency and industrial democracy.

(Miller 1989, 36)

These are powerful arguments. We acknowledged this point in Chapter 7 and it is worth re-emphasizing that the case for competitive markets has inevitably been strengthened by the recent collapse of most systems of centralized economic planning. Even so, the theory of open and competitive markets is one thing, and the actual operations of many real markets quite another. Before we turn to the case for market socialism, let us rehearse five important arguments against the hegemony of *the market* as it is often exercised in the contemporary world economy.

A first objection to *the market* is that even when its price-setting function is working reasonably well it fails properly to deal with externalities like pollution and with the provision of public goods like health and education. In regard to education, it is significant that such poor societies as Kerala in India, Sri Lanka and China have outperformed richer countries like Saudi Arabia and Iraq in part because they ploughed public monies into public education systems. Sri Lanka and China have also invested heavily in public healthcare provision and have seen life expectancy levels rise rapidly and well in advance of average or expected levels (Dasgupta 1993). It might also be relevant that life expectancy in England and Wales increased most at those times, after the two World Wars, when food rationing was in place and the provision of public healthcare services was expanded (Sen 1993).

A second objection to *the market* is that its price-setting function does not always work very well. A majority of real markets are neither open nor competitive. Prices are very often set by large firms (monopolies and oligopolies), as they are for example in the world food market where just five *merchants of grain* (agribusiness companies) control most of the world's grain trade. In most countries anti-trust legislation is either poorly developed and/ or poorly enforced. In many poorer countries, too, it is absurd to assume that a well-policed body of legislation exists to stop the exploitation of labour (including child-labour), or that most states finance an efficient health and safety inspectorate. Even in democratic developing countries some workers have few real rights (as opposed to paper rights), and in many authoritarian countries the quest for profits is allowed to ride roughshod over the legitimate rights and choices of workers. Many of the market reforms that have been introduced as part of structural adjustment and liberalization programmes have shown themselves to be anti-worker in practice, if not always in intent. It may be that there are few alternatives to structural adjustment measures in some indebted developing countries, and it may be that some of these reforms will pay off in the medium-term (as they appear to be doing in Argentina), but ten years into the banking-cum-debt crises there can be little doubt that today's market idolatry is hurting the poor and favouring the rich. The blunt fact is that real markets are historical–geographical creations that have built-in biases. Real markets tend to reproduce and amplify existing advantages of place, household and person. The abstract theory of open and competitive markets fails to take this properly into account.

A third argument against *the market* is that states and markets both fail on occasion, and there is now a body of empirical evidence that suggests that state interventions in the economies of low-income or transitional countries can be beneficial in the short-run. In the nineteenth century this view was associated with Friedrich List. List supported the use of tariff barriers to protect infant industries in agrarian economies. More recently the argument has been restated by many scholars working on the South Korean and Taiwanese 'miracles'. Robert Wade is not alone in arguing that import-substitution industrialization, properly managed by an efficient developing state, can lay the foundations for a later shift to export-orientated indus-trialization (Wade 1990). What matters is the quality of state interventions in the economy, a point Keynes made repeatedly in the first half of this century. Left-wing and right-wing ideologues have each erred in taking a 'two legs good, four legs bad' approach to either the private sector (the Left) or the public sector (the Right). Given that pure markets rarely exist in practice (save perhaps in financial services), it follows that arguments which oppose markets to states are both invalid and unhelpful. Their main function is ideological.

A fourth argument against *the market* is that an unregulated global market economy is heightening levels of risk in what Beck (1992) and Giddens (1990) have called 'radically modern capitalism'. Markets are good at dealing with risk after the event, as we have seen in the case of the 1980s' debt-cum-banking crises. New insurance mechanisms can be quickly tried and tested and the best of these mechanisms can be expected to survive. But free markets are less able to anticipate the nature of future risks, or to provide means for dealing with generalized risks that involve a measure of long-term damage or system irreversibility. Beck cites the threat to parts of the global environment in this last regard, and rightly so. The clear-felling of tropical rainforests might make sound commercial sense for some economic agents in the short-to-medium terms, but if this clear-felling is generalized across entire regions it can set in train a set of undesirable climatic changes that will be difficult to reverse in less than fifty or one hundred years. One answer to this threat is to internalize a set of environmental costs and benefits into the market 'equation' or process. Another way of dealing with such generalized risks is to accept that many such markets must be regulated by responsible governments or other public agencies.

A fifth argument against *the market* is, we believe, the most significant, because it is potentially the most radical and empowering. Hayek many times maintained that he was 'not a Conservative'. Quite rightly he saw that markets must guard against entrenched hierarchies of all kinds if they are to work properly. Truly open and competitive markets are not compatible with the existence of massive inequalities in income (and thus opportunity) that are the product of inheritance systems and powers of monopoly rather than hard work and personal initiative. Most socialists have put this case more bluntly still (Roemer 1988). In their view and in ours, it is a fiction to suppose that real men and women come to real markets in the late twentieth century with equal opportunities for participation, gain and loss. This is true neither within rich countries like the UK and the USA nor, manifestly, on a global basis. Real people are usually the prisoners, and more rarely the beneficiaries, of past histories and geographies: of their places of birth and the life-histories of their parents and grandparents and other kin. Of course, some people will escape their collective pasts-and-places, but this is not the point on which we should be focusing. Market ideologues hold out the possibility that anyone can be a winner in the lottery that is the global market-place, but they fail to point out that not everyone can be a winner at the same time, or that for every 'winner' there must be more 'losers'. This last point does not commit us to a simplistic zero-sum model of capitalist society – we can all be winners in absolute terms if entrepreneurship confers growth on a region or country – but it is a powerful riposte to the claims to justice as 'just deserts' that are built into the new ideology of market-access economics. The main objection to the market remains as before. Real people do not gain equal access to real markets and the rewards that most markets confer are, partly for this reason, unjust and hard to defend.

At one time proponents of markets as devices for decentralizing decision-making understood their limitations. Adam Smith, for example, thought they could work well only in alliance with social obligation. Self-interest was never enough. Another limitation was that filtering or trickling down (a favourite market metaphor for the benefits to all of an initial inequality in incomes) could only work if it did not involve consumption of the same goods. With the onset of mass consumption this is precisely what happened. Many people now aspire to consume (country houses, exotic holidays, fine wines) what previously was consumed by a minority. This often proves impossible because so many of these goods cannot be consumed without interfering with someone else's consumption. Wicksteed (1912) identified the central issue, later developed to great effect by Fred Hirsch (1976), when he insisted that though Napoleon might have liked to encourage the belief that each soldier carried in his knapsack a marshal's baton, it was obviously impossible that every soldier could become a marshal. The recent mania for the free-market is an example of hubris of Napoleonic proportions. It promises everyone a marshal's baton. It cannot deliver.

There is, however, a sting in the tail. We have noted already that markets do serve a useful if imperfect price-setting purpose, and recent history has taught us that state failures are every bit as likely as market failures. Strong doses of central planning have also tended to go hand in hand with an abuse of personal freedoms, even if, as in Maoist China, there is some evidence to suggest that positive freedoms (the right to eat, work and be housed) were increased more rapidly than in most non-socialist economies (Sen 1989). What we need to argue for today, we believe, is **fairer** real markets, and we can begin this task by turning the logic of open and competitive markets back upon those who have indulged in market triumphalism. 'How and in what ways', we need to ask, 'is the present global tendency to the formation of giant transnational corporations compatible with *the market?*' 'How and in what ways are tariff and non-tariff barriers in the West compatible with comparative advantage theories and the market?' 'How and in what ways can poorer regions and countries gain fair access to the *global market-place?*' (These arguments need not be transitive or symmetrical. Because we want to argue for managed markets – for the reasons set out above – we would happily argue for infant-industry supports in most transitional economies. Similarly, we would want to argue for public intervention in all economies in favour of investments in education, healthcare, infrastructure and so on. We only want to subvert part of the case for the market in its own terms; on many other occasions we would argue for particular social and economic projects on the basis of market failures, externalities and so on.) Finally (for our purposes here), we need to ask 'how and in what ways present systems of income and wealth transfer are compatible with ensuring a real equality of opportunity in, or access to, real markets?' Market socialists in the UK have suggested that all inter-generational transfers of wealth should be taxed severely, with the proceeds being used by the state to improve the life-chances of disadvantaged men and women entering the labour market (at age 16, 18 or whatever: Abell 1989).

We support this proposition and note that it can be justified by an appeal both to the proper workings of open and competitive markets and to social justice. We also support calls for greater industrial democracy in market-based economies and, as we said before, calls for greater international transfers of resources between countries and regions and calls for reforms of the global market-place. Needless to say, we are also in favour of legal protection for consumers and workers in markets and workplaces the world over. The problem with most structural adjustment programmes as they are currently conceived is that they enshrine rights for employers rather than employees. Workers need the protection of bodies like a beefed up International Labour Organization; a point that returns us to the discourse of internationalism–multilateralism. Citizens also need the protection offered to them as 'global individuals' by, for example, the International Court of Justice. The development of such supranational organizations is clearly at one with our general prospectus for a more reciprocal international order. If the future does lie with *the world market* in some form, it is important that market-places are everywhere democratized and subjected to certain common forms of political regulation.

CONCLUSION: DISCOURSE COUNTS

We are acutely aware that the foregoing discussion offers only a thumbnail sketch of a possible challenge to the new hegemony of transnational liberalism. Much work remains to be done. Nevertheless, we do strongly believe that discourse counts and that intellectuals have a responsibility to help fashion – and interrogate – new *representational spaces* that are resistant to some aspects of the emerging world order.

The discourse of the free-market has had important effects on the establishment of transnational liberalism as an incipient geopolitical order. The key problem for an alternative geopolitical discourse is to articulate its geographical vision of a world in which the market is at once tamed, decentralized and 'disestablished', and where empowered global citizens are able to challenge opposing elements of the present dynamic of globalization. To this end we have proposed the discourses of internationalism–multilateralism, of local oppositional imaginations and of market socialism, as three sets of vocabularies that challenge conventional attempts to master space by disempowering communities. If we were to recast these vocabularies, we would say that we are proposing a counter-hegemonic politics of empowerment that finds its roots in what Paul Hirst (1993) has called 'associational socialism' (combining market socialism and a pluralist state) and a still broader language of reciprocity. In geographical terms the meaning of this change in institutions and language would be a deprivatization and deterritorialization of space into *places* for securing human life against the depredations of coercive hierarchy. In reclaiming space for people this vision rejects the inevitability of the mastering of space by either states or markets.

We do not, of course, believe that it is only discourse that counts, or that words and actions can or should be easily separated. Certainly, it will be difficult in practice to challenge the current ascendancy of mobile capital, the world market, and the United States and its various allies. That said, the strangeness of the world after the Cold War testifies to the suddenness of effective discursive extinction. Surrounded by the debris of the old geopolitical order we find inspiration for challenging the cast of a new one.

REFERENCES

Abell, P. (1989) 'An equitarian market socialism', in J. Le Grand and S. Estrin (eds) *Market Socialism*, Oxford: Clarendon Press.

Abowd, J. M. and Freeman, R.B. (1990) 'The internationalization of the U.S. labor market', Working Paper, National Bureau of Economic Research, Cambridge MA.

Achebe, C. (1975) *Morning Yet on Creation Day*, London: Faber.

Agarwal, B. (1986) *Cold Hearths and Barren Slopes*, London: Zed.

Agnew, J. A. (1984) 'Devaluing place: "People prosperity" versus "place prosperity" and regional planning', *Society and Space* 2: 35–45.

—— (1987a) *Place and Politics: The Geographical Mediation of State and Society*, London: Allen and Unwin.

—— (1987b) *The United States in the World Economy: A Regional Geography*, Cambridge: Cambridge University Press.

—— (1989) 'The devaluation of place in social science', in J. A. Agnew and J. S. Duncan (eds) *The Power of Place: Bringing Together Geographical and Sociological Imaginations*, Boston: Unwin Hyman.

—— (1993) 'Representing space: space, scale and culture in social science', in J. Duncan and D. Ley (eds) *Place/Culture/Representation*, London: Routledge.

Agnew, J. A. and Corbridge, S. (1989) 'The new geopolitics: the dynamics of geopolitical disorder', in R. J. Johnston and P. J. Taylor (eds) *A World in Crisis? Geographical Perspectives*, 2nd edn, Oxford: Blackwell.

Aho, C. M. and Ostry, S. (1990) 'Regional trading blocs', in W. Brock and R. D. Harmat (eds) *The Global Economy: America's Role in the Decade Ahead*, New York: Norton.

Albert, B. (1992) 'Indian lands, environmental policy and military geopolitics in the development of the Brazilian Amazon: the case of the Yanomani', *Development and Change* 23: 35–70.

Altvater, E. (1993) *The Future of the Market: An Essay on the Regulation of Money and Nature After the Collapse of 'Actually Existing Socialism'*, London: Verso.

Amsden, A. (1989) *Asia's Next Giant*, Oxford: Oxford University Press.

Anderson, B. (1983) *Imagined Communities*, London: Verso.

Anderson, P. (1976/77) 'The antinomies of Antonio Gramsci', *New Left Review* 100: 5–78.

Andrews, J. (1992) 'A change of face: a survey of Taiwan', *Economist* 10 October: survey section.

Applegate, C. (1990) *A Nation of Provincials: The German Idea of Heimat*, Berkeley CA: University of California Press.

Arrighi, G. (1990) 'The three hegemonies of historical capitalism', *Review* 13: 365–408.

REFERENCES

—— (1991) 'World income inequalities and the future of socialism', *New Left Review* 189: 39–65.

Arrow, K. (1951) *Social Choice and Individual Values*, New York: Wiley.

Art, R. J. (1991) 'A defensible defense: America's grand strategy after the Cold War', *International Security* 15: 5–53.

Ash, T. G. (1990) *The Uses of Adversity: Essays on the Fate of Central Europe*, New York: Vintage.

Ashley, R.K. (1984) 'The poverty of neorealism', *International Organization* 38: 225–86.

—— (1988) 'Untying the sovereign state: a double reading of the anarchy problematique', *Millennium* 17: 227–62.

Augelli, E. and Murphy, C. N. (1993) 'Gramsci and international relations: a general perspective with examples from recent US policy toward the Third World', in S. Gill (ed.) *Gramsci, Historical Materialism and International Relations*, Cambridge: Cambridge University Press.

Bacon, K. H. (1993) 'Banks' declining role in economy worries Fed, may hurt firms', *Wall Street Journal* 9 July: A1, A5.

Bagby, L. M. J. (1994) 'The use and abuse of Thucydides in international relations', *International Organization* 48: 131–53.

Baker, C. (1981) 'Economic reorganization and the slump in South and Southeast Asia', *Comparative Studies in Society and History* 23: 325–49.

Bardhan, P. (1984) *The Political Economy of Development in India*, Oxford: Oxford University Press.

Barnouw, D. (1990) *Visible Spaces: Hannah Arendt and the German–Jewish Experience*, Baltimore: The Johns Hopkins University Press.

Barrett, M. (1991) *The Politics of Truth: From Marx to Foucault*, Stanford CA: Stanford University Press.

Barro, R. and Grossmann, H. (1976) *Money, Employment and Inflation*, Cambridge: Cambridge University Press.

Bassin, M. (1987) 'Race contra space: the conflict between German *Geopolitik* and National Socialism', *Political Geography Quarterly* 6: 115–34.

—— (1991) 'Russia between Europe and Asia: the ideological construction of geographical space', *Slavic Review* 50: 1–17.

—— (1993) 'Turner, Solov'ev, and the "frontier hypothesis": the nationalist significance of open spaces', *Journal of Modern History* 65: 473–511.

Bates, R. (1981) *Markets and States in Tropical Africa*, Berkeley CA: University of California Press.

—— (ed.) (1988) *Toward a Political Economy of Development*, Berkeley CA: University of California Press.

Bauer, P. (1972) *Dissent on Development*, London: Weidenfeld and Nicolson.

—— (1981) *Equality, the Third World and Economic Delusion*, London: Methuen.

—— (1991) *The Development Frontier: Essays in Applied Economics*, Hemel Hempstead: Harvester Wheatsheaf.

Bauman, Z. (1990) 'Effacing the face: on the social management of moral proximity', *Theory, Culture and Society* 7: 5–38.

Bayart, J.-F. (1993) *The State in Africa: The Politics of the Belly*, Harlow: Longman.

Bayoumi, T. and Eichengreen, B. (1992) 'Is there a conflict between EC enlargement and European monetary unification?', Working Paper, National Bureau of Economic Research, Cambridge MA.

Beck, U. (1992) *Risk Society: Towards a New Modernity*, London: Sage.

Becker, D. and Sklar, R. (eds) (1987) *Postimperialism: International Capitalism and Development in the Late Twentieth Century*, Boulder CO: Lynne Rienner.

Beenstock, M. (1984) *The World Economy in Transition*, London: George Allen and Unwin.

Beitz, C. (1991) 'Sovereignty and morality in international affairs', in D. Held (ed.) *Political Theory Today*, Cambridge: Polity.

Benedick, R. (1991) *Ozone Diplomacy: New Directions in Safeguarding the Planet*, Cambridge MA: Harvard University Press.

Benn, S. I. (1967) 'State', *Encyclopedia of Philosophy*, Volume VIII: 6–11, New York: Collier-Macmillan.

Berger, T. U. (1993) 'From sword to chrysanthemum: Japan's culture of anti-militarism', *International Security* 17: 119–50.

Bergsten, C. F. (1982) 'The United States and the world economy', *Annals of the American Academy of Political and Social Science* 460: 11–20.

Bergsten, C. F. and Cline, W. R. (1987) *The United States–Japan Economic Problem*, Rev. edn, Washington DC: Institute for International Economics.

Bhaskar, R. (1979) *The Possibility of Naturalism*, Brighton: Wheatsheaf.

Biersteker, T. (1992) 'The "triumph" of neoclassical economics in the developing world: policy convergence and the bases of government in the international economic order', in J. Rosenau and E.-O. Czempiel (eds) *Governance Without Government: Order and Change in World Politics*, Cambridge: Cambridge University Press.

—— (1993) 'Evolving perspectives on international political economy: twentieth-century contexts and discontinuities', *International Political Science Review* 14: 7–33.

Black, R. (1991) 'Refugees and displaced persons', *Progress in Human Geography* 15: 281–98.

Bourdieu, P. (1977) *Outline of a Theory of Practice*, Cambridge: Cambridge University Press.

Bowles, S., Gordon, D. and Weisskopf, T. E. (1983) *Beyond the Wasteland*, Garden City NY: Anchor-Doubleday.

Bowman, M. J. and Harris, J. J. (1984) *Multilateral Treaties: Index and Current Status*, London: Butterworth.

Brett, E. (1985) *The World Economy Since the War: The Politics of Uneven Development*, London: Macmillan.

Brittan, S. (1975) 'The economic contradictions of democracy', *British Journal of Political Science* 5: 129–59.

Brown, L. (1992) 'Ecology and world security', *The World Today* 48: 51–5.

Bruno, M. and Sachs, J. (1985) *Economics of Worldwide Stagflation*, Oxford: Blackwell.

Bryant, R. (1991) 'Political ecology: an emerging research agenda in Third World studies', *Political Geography Quarterly* 11: 12–36.

Bull, H. (1977) *The Anarchical Society: A Study of Order in World Politics*, New York: Columbia University Press.

Burleigh, M. (1988) *Germany Turns Eastward: A Study of Ostforschung in the Third Reich*, Cambridge: Cambridge University Press.

Cafruny, A. W. (1990) 'A Gramscian concept of declining hegemony: stages of U.S. power and the evolution of international economic relations', in D. P. Rapkin (ed.) *World Leadership and Hegemony*, Boulder CO: Lynne Rienner.

Calder, K. (1988) 'Japanese foreign economic policy formation: explaining the reactive state', *World Politics* 40: 517–41.

Calleo, D. P. (1982) *The Imperious Economy*, Cambridge MA: Harvard University Press.

—— (1987) *Beyond American Hegemony: The Future of the Western Alliance*, New York: Basic Books.

—— (1990) 'The US in the 1960s: hegemon in decline?', in M. Mann (ed.) *The Rise and Decline of the Nation State*, Oxford: Blackwell.

—— (1992) *The Bankrupting of America: How the Federal Budget is Impoverishing the Nation*, New York: Morrow.

Camilleri, J. and Falk, J. (1992) *The End of Sovereignty? The Politics of a Shrinking and Fragmenting World*, Aldershot: Edward Elgar.

Campbell, B. (1993) *Goliath: Britain's Dangerous Places*, London: Vintage.

Campbell, D. (1992) *Writing Security: United States Foreign Policy and the Politics of Identity*, Minneapolis MN: University of Minnesota Press.

Caporaso, J. A. (1992) 'International relations theory and multilateralism: the search for foundations', *International Organization* 46: 599–632.

Carr, E. H. (1939) *The 20 Years' Crisis, 1919–1939: An Introduction to the Study of International Relations*, London: Macmillan.

Cassen, R. *et al.* (1994) *Does Aid Work?: Report to an Intergovernmental Task Force*, 2nd edn, Oxford: Clarendon Press.

Castells, M. (1980) *The Economic Crisis and American Society*, Princeton NJ: Princeton University Press.

—— (1988) *The Informational City*, Oxford: Blackwell.

Chomsky, N. (1992) *Deterring Democracy*, London: Verso.

Clark, I. (1991) *The Hierarchy of States: Reform and Resistance in the International Order*, Cambridge: Cambridge University Press.

Claude, I. L. (1988) *States and the Global System: Politics, Law and Organization*, London: Macmillan.

Claude, R. P. and Davis, D. R. (1988) 'Political legitimacy at risk: the emergence of human rights in international politics', Paper presented at the Fourteenth World Congress of the International Political Science Association, Washington DC, September.

Cline, W. R. (1993) 'Macroeconomics and the US–Japan trade imbalance', *International Economic Insights* July/August: 5–8.

Cockburn, A. (1983) *The Threat: Inside the Soviet Military Machine*, New York: Random House.

Colclough, C. (1991) 'Stucturalism versus neo-liberalism: an introduction', in C. Colclough and J. Manor (eds) *States or Markets: Neo-Liberalism and the Development Policy Debate*, Oxford: Clarendon Press.

Collins, R. (1986) 'Modern technology and geopolitics', in R. Collins, *Weberian Sociological Theory*, New York: Cambridge University Press.

Collins, S. (1989) *From Divine Cosmos to Sovereign State: An Intellectual History of Consciousness and the Idea of Order in Renaissance England*, New York: Oxford University Press.

Connolly, W. E. (1988) *Political Theory and Modernity*, Oxford: Blackwell.

Conybeare, J. A. C. (1984) 'Public goods, prisoners' dilemmas and the international political economy', *International Studies Quarterly* 28: 5–22.

Cooper, R. (1991) 'Cross-border savings flows and capital mobility in the G7 economies', Bank of England Discussion papers, Number 54.

Corbridge, S. E. (1986) *Capitalist World Development: A Critique of Radical Development Geography*, London: Macmillan.

—— (1991) 'Marxisms, modernities, and moralities: development praxis and the claims of distant strangers', Unpublished paper, Department of Geography, University of Cambridge.

—— (1993) *Debt and Development*, Oxford: Blackwell.

—— (1994a) 'Plausible worlds: Friedman, Keynes and the geography of inflation', in S. Corbridge, R. Martin and N. Thrift (eds) *Money, Power and Space*, Oxford: Blackwell.

—— (1994b) 'Maximizing entropy? New geopolitical orders and the internationalization of business', in G. Demko and W. Wood (eds) *New World Order: Geopolitical Perspectives*, Boulder CO: Westview.

Corbridge, S. and Agnew, J. (1991) 'The US trade and budget deficits in global perspective: an essay in geopolitical-economy', *Society and Space* 9: 71–90.

Corbridge, S. and Thrift, N. (1994) 'Money, power and space: introduction and overview,' in S. Corbridge, R. Martin and N. Thrift (eds) *Money, Power and Space*, Oxford: Blackwell.

Cowhey, P. F. (1993) 'Domestic institutions and the credibility of international commitments: Japan and the United States', *International Organization* 47: 299–326.

Cowhey, P. F. and Aronson, J. D. (1993) *Managing the World Economy: The Consequences of Corporate Alliances*, New York: Council on Foreign Relations Press.

Cowles, M. G. (1993) 'The rise of the European multinational', *International Economic Insights* 4: 15–18.

Cox, K. R. and Mair, A. (1988) 'Locality and community in the politics of local economic development', *Annals of the Association of American Geographers* 78: 307–25.

Cox, M. (1990) 'From the Truman doctrine to the second superpower detente: the rise and fall of the Cold War', *Journal of Peace Research* 27: 25–41.

Cox, R. W. (1981) 'Social forces, states, and world orders: beyond international relations theory', *Millennium* 10: 126–55.

—— (1987) *Production, Power, and World Order*, New York: Columbia University Press.

—— (1992) 'Towards a post-hegemonic conceptualization of world order: reflections on the relevancy of Ibn Khaldun', in J. N. Rosenau and E.-N. Czempiel (eds) *Governance Without Government: Order and Change in World Politics*, Cambridge: Cambridge University Press.

Crane, G. T. and Amawi, A. (1991) *The Theoretical Evolution of International Political Economy*, New York: Oxford University Press.

Crockatt, R. (1993) 'Theories of stability and the end of the Cold War', in M. Bowker and R. Brown (eds) *From Cold War to Collapse; Theory and World Politics in the 1980s*, Cambridge: Cambridge University Press.

Cumings, B. (1984) 'The origins and development of the Northeast Asian political economy: industrial sectors, product cycles, and political consequences', *International Organization* 38: 1–40.

—— (1992) 'The wicked witch of the West is dead. Long live the wicked witch of the East', in M. J. Hogan (ed.) *The End of the Cold War: Its Meaning and Implications*, Cambridge: Cambridge University Press.

Cutler, D. M. and Katz, L. F. (1992) 'Rising inequality? Changes in the distribution of income and consumption in the 1980's', *American Economic Review* 82: 546–51.

Dalby, S. (1988) 'Geopolitical discourse: the Soviet Union as other', *Alternatives* 13: 415–42.

—— (1990) *Creating the Second Cold War*, New York: Guilford.

—— (1991) 'Rethinking security: ambiguities in policy and theory', Occasional Paper, International Studies, Simon Fraser University, Burnaby BC, Canada.

—— (1992) 'Ecopolitical discourse; "environmental security" and political geography', *Progress in Human Geography* 16: 503–22.

Dallek, R. (1983) *The American Style of Foreign Policy: Cultural Politics and Foreign Affairs*, New York: Mentor.

Daly, M. and Logan, M. (1989) *The Brittle Rim: Finance, Business and the Pacific Region*, London: Penguin.

232

Daniels, P. (1993) *Service Industries in the World Economy*, Oxford: Blackwell.

Darwin, C. (1839) *Journal of Researches into the Geology and Natural History of the Various Countries Visited by the H.M.S. Beagle*, London: Henry Colburn.

Dasgupta, P. (1993) *An Inquiry into Well-Being and Destitution*, Oxford: Oxford University Press.

Davis, M. (1985) 'Reaganomics' magical mystery tour', *New Left Review* 149: 45–65.

—— (1990) *City of Quartz*, London: Verso.

Dehio, L. (1962) *The Precarious Balance: The Politics of Power in Europe, 1494–1945*, London: Chatto and Windus.

Deibel, T. L. (1992) 'Strategies before containment: patterns for the future', *International Security* 16: 79–108.

deLemos, H. (1990) 'Amazonia: in defence of Brazil's sovereignty', *The Fletcher Forum of World Affairs* 14: 301–12.

Denitch, B. (1993) 'Learning from the death of Yugoslavia: nationalism and democracy', *Social Text* 34: 3–16.

Denoon, D. B. H. (1993) *Real Reciprocity: Balancing U.S. Economic and Security Policy in the Pacific Basin*, New York: Council on Foreign Relations Press.

Der Derian, J. (1990) 'The (s)pace of international relations: simulation, surveillance, and speed', *International Studies Quarterly* 34: 295–310.

Dessler, D. (1989) 'What's at stake in the agent–structure debate?', *International Organization* 43: 441–73.

Deudney, D. (1990) 'The case against linking environmental degradation and national security', *Millennium* 19: 461–76.

—— (1993) 'Bringing nature back in: concepts, problems and trends in physiopolitical theory from the Greeks to the greenhouse', Paper presented at the American Political Science Association, Washington DC, September.

Deudney, D. and Ikenberry, G. J. (1991) 'Soviet reform and the end of the Cold War: explaining large-scale historical change', *Review of International Studies* 17: 225–50.

—— (1991/2) 'The international sources of Soviet change', *International Security* 16: 74–118.

Dicken, P. (1992) *Global Shift: The Internationalization of Economic Activity*, 2nd edn, New York and London: Guilford Press and Paul Chapman.

Dornbusch, R. and Edwards, S. (eds) (1991) *The Macroeconomics of Populism in Latin America*, Cambridge MA: MIT Press.

Doty, R. (1993) 'Sovereignty and national identity: constructing the nation', Unpublished paper, Department of Political Science, Arizona State University, Tempe AZ.

Duncan, J. (1993) 'Sites of representation: place, time and the discourse of the other', in J. Duncan and D. Ley (eds) *Place/Culture/Representation*, London: Routledge.

Dunn, J. (1979) *Western Political Theory in the Face of the Future*, Cambridge: Cambridge University Press.

Eco, U. (1987) *Travels in Hyperreality*, San Diego: Harcourt Brace Jovanovich.

Economist (1991) 'The future League of Nations', 24 August: 56.

—— (1992) 'Europe's hard core', 21 November: 77.

—— (1993a) 'Germany and France: a matter of business', 6 March: 19–22.

—— (1993b) 'China belongs to me', 29 May: 13–14.

Edmonds, R. (1983) *Soviet Foreign Policy: The Brezhnev Years*, New York: Oxford University Press.

Edwards, R. and Garonna, P. (1991) *The Forgotten Link: Labor's Stake in International Economic Cooperation*, Savage MD: Rowman and Littlefield.

Eichengreen, B. (1988) 'Hegemonic stability theories of the international monetary system', *Brookings Papers on International Economics* 54.

Eichengreen, B. and Lindert, P. (eds) (1989) *The International Debt Crisis in Historical Perspective*, Cambridge MA: MIT Press.

Eller, J. D. and Coughlan, R. M. (1993) 'The poverty of primordialism: the demystification of ethnic attachments', *Ethnic and Racial Studies* 16: 183–202.

Encarnation, D. J. and Mason, M. (1990) 'Neither MITI nor America: the political economy of capital liberalization in Japan', *International Organization* 44: 25–54.

Esposito, J. L. (1992) *The Islamic Threat: Myth or Reality?*, New York: Oxford University Press.

Esteva, G. (1992) 'Development', in W. Sachs (ed.) *The Development Dictionary*, London: Zed Books.

Estrin, S. and Le Grand, J. (1989) 'Market socialism', in J. Le Grand and S. Estrin (eds) *Market Socialism*, Oxford: Clarendon Press.

Evans, R. J. (1989) *In Hitler's Shadow: West German Historians and the Attempt to Escape from the Nazi Past*, New York: Pantheon.

Fabian, J. (1983) *Time and the Other: How Anthropology Makes its Object*, New York: Columbia University Press.

Fagan, R. (1990) 'Elders IXL Ltd.: finance capital and the geography of corporate restructuring', *Environment and Planning A* 22: 647–66.

Fairgreave, J. (1932) *Geography of World Power*, London: University of London Press.

Faux, J. (1993) 'The crumbling case for NAFTA', *Dissent* Summer: 309–15.

Feldman, G. D. (1994) *The Great Disorder: Politics, Economics, and Society in the German Inflation, 1914–1924*, New York: Oxford University Press.

Feldstein, M. and Horioka, C. (1980) 'Domestic savings and international capital flows', *Economic Journal* 91: 17–31.

Fieleke, N. S. (1992) 'One trading world, or many? the issue of regional trading blocs', *New England Economic Review* May/June: 3–20.

Fischer, M. (1992) 'Feudal Europe, 800–1300: communal discourse and conflictual practice', *International Organization* 46: 427–66.

—— (1993) 'On context, facts, and norms: reply to Hall and Kratochwil', *International Organization* 47: 493–500.

Fox-Genovese, E. (1991) *Feminism Without Illusions: A Critique of Individualism*, Chapel Hill: University of North Carolina Press.

Frankel, J. (1991) 'Is a yen bloc forming in Pacific Asia?', in R. O'Brien (ed.) *Finance and the International Economy*, New York: Oxford University Press.

Friedberg, A. L. (1988) *The Weary Titan: Britain and the Experience of Relative Decline, 1895–1905*, Princeton NJ: Princeton University Press.

—— (1992) 'Why didn't the United States become a garrison state?', *International Security* 16: 109–42.

Friedman, B. M. (1987) 'Long-run costs of US fiscal policy: the international dimension', Unpublished paper, Department of Economics, Harvard University.

—— (1988) *Day of Reckoning: The Consequences of American Economic Policy*, New York: Vintage.

—— (1992) 'Learning from the Reagan deficits', *American Economic Review* 82: 299–304.

Friedman, G. and LeBard, M. (1991) *The Coming War with Japan*, New York: St. Martin's Press.

Friedman, M. (1982) *Capitalism and Freedom*, Rev. edn, Chicago: University of Chicago Press.

Friedman, M. and Schwartz, A. (1963) *A Monetary History of the United States, 1867–1960*, Princeton NJ: Princeton University Press.

—— (1982) *Monetary Trends in the United States and the United Kingdom: Their*

Relation to Income, Prices, and Interest Rates, 1867–1975, Chicago: University of Chicago Press.

Friedrich, P. (1989) 'Language, ideology, and political economy', *American Anthropologist* 91: 295–312.

Frug, G. E. (1980) 'The city as a legal concept', *Harvard Law Review* 93: 1059–154.

Fukuyama, F. (1989) 'The end of history', *The National Interest* 16: 3–16.

Gaddis, J. L. (1982) *Strategies of Containment*, New York: Oxford University Press.

—— (1987) 'Introduction: the evolution of containment', in T. L. Deibel and J. L. Gaddis (eds) *Containing the Soviet Union*, Washington DC: Pergamon-Brassey's.

Gamble, A. (1988) *The Free Economy and the Strong State: The Politics of Thatcherism*, London: Macmillan.

Garrett, G. (1992) 'International cooperation and institutional choice; the European Community's internal market', *International Organization* 46: 533–60.

Garrett, G. and Lange, P. (1991) 'Political responses to interdependence: what's "left" for the left?' *International Organization* 45: 539–64.

Garrison, R. and Kirzner, I. (1989) 'Friedrich August von Hayek', in J. Eatwell, M. Milgate and P. Newman (eds) *The New Palgrave: The Invisible Hand*, London: Macmillan.

Garst, D. (1989) 'Thucydides and neorealism', *International Studies Quarterly* 33: 3–27.

GATT (1990) *Trade Policy Review – Japan*, Geneva: General Agreement on Tariffs and Trade.

Gervasi, T. (1988) *Soviet Military Power: The Pentagon's Propaganda Document Annotated and Corrected*, New York: Vintage.

Giddens, A. (1984) *The Nation-State and Violence*, London: Macmillan.

—— (1990) *The Consequences of Modernity*, Cambridge: Polity.

Giersch, H. (1992) *The Fading Miracle: Four Decades of Market Economy in Germany*, Cambridge: Cambridge University Press.

Gill, S. (1990) *American Hegemony and the Trilateral Commission*, Cambridge: Cambridge University Press.

—— (ed.) (1993) *Gramsci, Historical Materialism and International Relations*, Cambridge: Cambridge University Press.

Gill, S. and Law, D. (1988) *The Global Political Economy*, Baltimore: The Johns Hopkins University Press.

Gilman, S. L. (1992) 'Plague in Germany, 1939/1989; cultural images of race, space, and disease', in A. Parker, M. Russo, D. Summer and P. Yaeger (eds) *Nationalisms and Sexualities*, London: Routledge.

Gilpin, R. (1981) *War and Change in World Politics*, Cambridge: Cambridge University Press.

—— (1987) *The Political Economy of International Relations*, Princeton: Princeton University Press.

Glaser, C. L. (1993) 'Why NATO is still best: future security arrangements for Europe', *International Security* 18: 5–50.

Gleick, P. (1989) 'The implications of global climatic change for international security', *Climatic Change* 15: 309–25.

Goertz, G. and Diehl, P. F. (1992) *Territorial Changes and International Conflict*, London: Routledge.

Goldgeier, J. M. and McFaul, M. (1992) 'A tale of two worlds: core and periphery in the post-Cold War era', *International Organization* 46: 467–91.

Goodman, J. B. (1992) *Monetary Sovereignty: The Politics of Central Banking in Western Europe*, Ithaca NY: Cornell University Press.

Gorbachev, M. S. (1987/8) 'The reality and guarantees of a secure world', *Foreign Broadcast Information Service* SOV-87–180, 17 September.

Gordon, D. (1988) 'The global economy: new edifice or crumbling foundations?' *New Left Review* 168: 24–64.

Grant, R. (1993) 'Trading blocs or trading blows? The macroeconomic geography of US and Japanese trade policies', *Environment and Planning A* 25: 273–91.

Gray, C. (1988) *The Geopolitics of Superpower*, Lexington KY: University Press of Kentucky.

Greengrass, M. (ed.) (1991) *Conquest and Coalescence: The Shaping of the State in Early Modern Europe*, London: Arnold.

Gregory, R. (1978) 'The domino theory', in A. DeConde (ed.) *Encyclopedia of American Foreign Policy Vol. 1*, New York: Charles Scribner's Sons.

Greider, W. (1987) *Secrets of the Temple: How the Federal Reserve Runs the Country*, New York: Simon and Schuster.

Grimwade, N. (1989) *International Trade: New Patterns of Trade, Production and Investment*, Basingstoke: Macmillan.

Grunberg, I. (1990) 'Exploring the "myth" of hegemonic stability', *International Organization* 44: 431–77.

Guha, R. (1989) 'Dominance without hegemony and its historiography', in R. Guha (ed.) *Subaltern Studies* 6: 210–309.

Gupta, A. and Ferguson, J. (1992) 'Beyond "culture": space, identity, and the politics of difference', *Cultural Anthropology* 7: 6–23.

Habermas, J. (1987) *The Philosophical Discourse of Modernity*, Cambridge: Polity Press.

Hall, P. (ed.) (1989) *The Political Power of Economic Ideas: Keynesianism Across Nations*, Princeton NJ: Princeton University Press.

Hall, R. B. and Kratochwil, F. V. (1993) 'Medieval tales: neorealist "science" and the abuse of history', *International Organization*, 47: 479–91.

Hansson, K. (1951) 'A general theory of the system of multilateral trade', *American Economic Review* 42: 59–88.

Harries, M. and Harries, S. (1987) *Sheathing the Sword: The Demilitarization of Postwar Japan*, New York: Macmillan.

Harris, R. B. (1976) *Authority: A Philosophical Analysis*, University AL: University of Alabama Press.

Hartman, G. (ed.) (1986) *Bitburg in Moral and Political Perspective*, Bloomington IN: Indiana University Press.

Harvey, D. (1982) *The Limits to Capital*, Oxford: Blackwell.

—— (1989) *The Urban Experience*, Oxford: Blackwell.

—— (1990) *The Condition of Postmodernity*, Oxford: Blackwell.

Hawthorn, G. (1991) '"Waiting for a text?" Comparing Third World politics', in J. Manor (ed.) *Rethinking Third World Politics*, Harlow: Longman.

—— (1993a) 'Sub-Saharan Africa', in D. Held (ed.) *Prospects for Democracy*, Cambridge: Polity.

—— (1993b) 'Listen to the women: review of Dasgupta', *London Review of Books* 21 October: 18–19.

Hay, D. (1968) *The Age of the Renaissance*, London: Thames and Hudson.

Hayek, F. (1944) *The Road to Serfdom*, Chicago: University of Chicago Press.

—— (1960) *The Constitution of Liberty*, London: Routledge and Kegan Paul.

Hecht, S. and Cockburn, A. (1990) *The Fate of the Forest: Developers, Destroyers and Defenders of the Amazon*, London: Verso.

Hegel, G. W. F. [1921] (1942) *Philosophy of Right*, Oxford: Oxford University Press.

—— (1931) *The Phenomenology of Mind*, 2nd edn, New York: Humanities Press.

Heidegger, M. (1959) *An Introduction to Metaphysics*, New Haven CT: Yale University Press.

Heller, J. (1962) *Catch-22*, New York: Dell Books.

Helms, M. (1988) *Ulysses' Sail: An Ethnographic Odyssey of Power, Knowledge, and Geographical Distance*, Princeton NJ: Princeton University Press.

Henderson-Sellars, A. (1991) 'Policy advice on greenhouse induced climatic change: the scientists' dilemma', *Progress in Physical Geography* 15: 53–70.

Henrikson, A. (1991) 'Mental maps', in M. J. Hogan and T. G. Paterson (eds) *Explaining the History of American Foreign Relations*, Cambridge: Cambridge University Press.

Hepworth, M. (1989) *Geography of the Information Economy*, London: Belhaven.

Herz, J. H. (1957) 'The rise and demise of the territorial state', *World Politics* 9: 473–93.

Hewitt, K. (1983) 'Place annihilation: area bombing and the fate of urban places', *Annals of the Association of American Geographers* 73: 257–84.

Hinsley, F. H. (1966) *Sovereignty*, New York: Basic Books.

Hirsch, F. (1976) *Social Limits to Growth*, Cambridge MA: Harvard University Press.

Hirst, P. (1993) *Associative Democracy*, Cambridge: Polity.

Hobson, J. A. (1904) *Imperialism: A Study*, 2nd edn, London: Allen and Unwin.

Hodgson, G. (1993) 'Grand illusion, the failure of European consciousness', *World Policy Journal* 10, 2: 13–18.

Holdar, S. (1992) 'The ideal state and the power of geography: the life and work of Rudolf Kjellen', *Political Geography* 11: 307–23.

Holsti, K. J. (1992) 'Governance without government: polyarchy in nineteenth-century European international politics', in J. N. Rosenau and E.-N. Czempiel (eds) *Governance Without Government: Order and Change in World Politics*, Cambridge: Cambridge University Press.

Holzman, F. D. (1989) 'Politics and guesswork: CIA and DIA estimates of Soviet military spending', *International Security* 101–31.

Homer-Dixon, T. (1991) 'On the threshold: environmental changes as causes of acute conflict', *International Security* 16: 76–116.

Homer-Dixon, T. F., Boutwell, J. H. and Rathjens, G. W. (1993) 'Environmental change and violent conflict', *Scientific American* February: 38–45.

Hoogvelt, A. (1982) *The Third World in Global Development*, London: Macmillan.

Hosking, G. (1990) *The Awakening of the Soviet Union*, Cambridge MA: Harvard University Press.

Hout, W. (1993) *Capitalism and the Third World: Development, Dependence and the World System*, Aldershot: Edward Elgar.

Huntington, S. P. (1993a) 'Why international primacy matters', *International Security* 17: 68–83.

—— (1993b) 'The clash of civilizations?', *Foreign Affairs* 72: 22–49.

Hymer, S. (1975) 'The multinational corporation and the law of uneven development,' in H. Radice (ed.) *International Firms and Modern Imperialism*, London: Penguin.

Ibn Khaldun (1967) *Muqaddimah; An Introduction to History*, trans K. Rosenthal, Princeton NJ: Princeton University Press.

Ikenberry, G. J. (1986) 'The irony of state strength: comparative responses to the oil shocks of the 1970s', *International Organization* 40: 105–37.

Ikenberry, G. J. and Kupchan, C. A. (1990) 'The legitimation of hegemonic power', in D. Rapkin (ed.) *World Leadership and Hegemony*, Boulder CO: Lynne Rienner.

Iklé, F. C. and Wohlstetter, A. (1988) *Discriminate Deterrence*, Report of the Commission on Integrated Long-Term Strategy, Washington D.C.: Government Printing Office.

Inayatullah, N. (1993) 'Theories of spontaneous disorder; generating intentions in Waltz, Gilpin, and Adam Smith', Unpublished paper, Maxwell School, Syracuse University NY.

—— (1994) 'Beyond the sovereignty dilemma: international society, global division of labor and Third World states', *International Organization*, forthcoming.

Inayatullah, N. and Rupert, M. E. (1993) 'Hobbes, Smith and the problem of mixed ontologies in neorealist international political economy', in S. Rosow (ed.) *The Global Economy as Political Space*, Boulder CO: Lynne Rienner.

India Today (1993) 'Interview with Nagarajan Vittal,' International Edition, 31 August: 59.

Inoguchi, T. (1988/9) 'Four Japanese scenarios for the future', *International Affairs* January: 15–28.

Ito, T. (1992) 'Political economy of the yen bloc in Asia', Paper presented at Conference on Pacific Trade, September, Washington DC.

Jackson, R. H. (1990) *Quasi-states: Sovereignty, International Relations and the Third World*, Cambridge: Cambridge University Press.

Jalée, P. (1969) *The Third World in the World Economy*, London: Monthly Review Press.

James, S. C. and Lake, D. A. (1989) 'The second face of hegemony: Britain's repeal of the Corn Laws and the American Walker Tariff of 1846', *International Organization* 43: 1–29.

Jervis, R. (1985) 'From balance to concert: a study of international security co-operation', *World Politics* 38: 58–79.

—— (1993) 'International primacy: is the game worth the candle?', *International Security* 17: 52–67.

Johnson, C. (1982) *MITI and the Japanese Economic Miracle: The Growth of Industrial Policy, 1925–1975*, Stanford CA: Stanford University Press.

—— (1987) 'How to think about economic competition from Japan', in K. Pyle (ed.) *The Trade Crisis: How Will Japan Respond?* Seattle WA: Society for Japanese Studies.

Johnson, L. M. (1993) *Thucydides, Hobbes and the Interpretation of Realism*, DeKalb IL: Northern Illinois University Press.

Johnston, R. J. (1993) 'The rise and decline of the corporate-welfare state: a comparative analysis in global context', in P. J. Taylor (ed.) *The Political Geography of the Twentieth Century*, London: Belhaven Press.

Jones, R. S. (1988) 'The economic implications of Japan's aging population', *Asian Survey* 28: 17–31.

Jones, R. S., King, R. E. and Klein, M. (1993) 'Economic integration between Hong Kong, Taiwan and the coastal provinces of China', *OECD Economic Studies*, 20: 115–44.

Jowitt, K. (1992) *New World Disorder: The Leninist Extinction*, Berkeley CA: University of California Press.

Judy, R. W. and Clough, V. L. (1989) *The Information Age and Soviet Society*, Indianapolis IN: Bobbs-Merrill.

Julius, D. (1990) *Global Companies and Public Policy: The Growing Challenge of Foreign Direct Investment*, New York: Council on Foreign Relations Press.

Kahler, M. (1992) 'Multilateralism with small and large numbers', *International Organization* 46: 681–708.

Kaldor, M. (1993) 'Le radici della guerra', *L'Indice dei Libri del Mese* giugno: 47–50.

Katzenstein, P. J. and Okawara, N. (1993) 'Japan's national security: norms, structures, and policies', *International Security* 17: 84–118.

Kaysen, C. (1990) 'Is war obsolete?', *International Security* 14: 60–72.

Keal, P. (1984) *Unspoken Rules and Superpower Dominance*, New York: St. Martin's Press.

Kearns, G. (1993) 'Prologue: *fin de siècle* geopolitics: Mackinder, Hobson and theories of global closure', in P. J. Taylor (ed.) *The Political Geography of the Twentieth Century*, London: Belhaven Press.

Kegley, C. W. Jr. and Wittkopf, E. R. (1993) *World Politics: Trends and Transformation*, 4th edn, New York: St. Martin's Press.

Kennan, G. [Mr. X] (1947) 'The sources of Soviet conduct', *Foreign Affairs* 25: 566–82.

Kennedy, P. (1987) *The Rise and Fall of the Great Powers: Economic Change and Military Conflict from 1500 to 2000*, New York: Random House.

—— (1993) *Preparing for the Twenty-First Century*, New York: Random House.

Keohane, R. O. (1984) *After Hegemony: Cooperation and Discord in the World Political Economy*, Princeton NJ: Princeton University Press.

Kern, S. (1983) *The Culture of Time and Space, 1880–1918*, Cambridge MA: Harvard University Press.

Keynes, J. M. (1936) *The General Theory of Employment, Interest and Money*, London: Macmillan.

—— (1971) *The Economic Consequences of the Peace*, London: Macmillan.

—— (1972) *Essays in Persuasion*, London: Macmillan.

Kindleberger, C.P. (1976) 'Systems of economic organizations', in D. P. Calleo (ed.) *Money and the Coming World Order*, New York: New York University Press.

—— (1986) 'Hierarchy versus inertial cooperation', *International Organization* 40: 841–8.

King, A. (1990) *Global Cities: Postimperialism and the Internationalization of London*, London: Routledge.

Kissinger, H. (1994) *Diplomacy*, New York: Simon and Schuster.

Kitschelt, H. (1991) 'Industrial governance structures, innovation strategies, and the case of Japan: sectoral or cross-national comparative analysis?', *International Organization* 45: 453–93.

Klare, M. (1993) 'The next great arms race', *Foreign Affairs* 72, 3: 136–52.

Knox, P. and Agnew, J. (1994) *The Geography of the World Economy*, 2nd edn, London: Edward Arnold.

Komorov, B. (1980) *The Destruction of Nature in the Soviet Union*, London: Pluto Press.

Kontorovich, V. (1992) 'Technological progress and research and development', in M. Ellman and V. Kontorovich (eds) *The Disintegration of the Soviet Economic System*, London: Routledge.

Kratochwil, F. (1989) *Rules, Norms and Decisions*, Cambridge: Cambridge University Press.

Kristof, L. (1968) 'The Russian image of Russia', in C. A. Fisher (ed.) *Essays in Political Geography*, London: Methuen.

Krueger, A. (1974) 'The political economy of the rent seeking society', *American Economic Review* 64: 291–303.

—— (1985) 'The importance of general policies to promote economic growth', *The World Economy* 8: 93–108.

—— (1992) 'Global trade prospects for the developing countries', *The World Economy* 15: 457–74.

—— (1993) *Economic Policies at Cross-Purposes: The United States and Developing Countries*, Washington DC: The Brookings Institution.

Krugman, P. R. (1992) 'The right, the rich, and the facts: deconstructing the income distribution debate', *The American Prospect* 11, Fall: 19–31.

—— (1994) *Peddling Prosperity: Economic Sense and Nonsense in the Age of Diminished Expectations*, New York: Norton.

Krugman, P. R. and Lawrence, R. Z. (1994) 'Trade, jobs, and wages', *Scientific American*, 270, April: 44–9.

Kuhl, S. (1994) *The Nazi Connection: Eugenics, American Racism, and German National Socialism*, New York: Oxford University Press.

Kume, I. (1988) 'Changing relations among the government, labor, and business in Japan after the oil crisis', *International Organization* 42: 659–87.

Lal, D. (1983) *The Poverty of 'Development Economics'*, London: Institute of Economic Affairs.

Lange, P. (1993) 'Maastricht and the social protocol: why did they do it?', *Politics and Society* 21: 5–36.

Lapidus, I. and Zaslavsky, V., with Goldman, P. (1992) *From Union to Commonwealth: Nationalism and Separatism in the Soviet Republics*, Cambridge: Cambridge University Press.

Laqueur, W. (1993) *Black Hundred: The Rise of the Extreme Right in Russia*, New York: Harper Collins.

Lash, S. and Urry, J. (1994) *Economies of Signs and Space*, London: Sage Publications.

Latham, A. J. H. (1978) *The International Economy and the Undeveloped World*, London: Croom Helm.

Lawson, N. (1991) *The View From Number 11*, London: Hutchinson.

Layne, C. (1993) 'The unipolar illusion: why new Great Powers will arise', *International Security* 17: 5–51.

Lefebvre, H. (1991) *The Production of Space*, Cambridge, MA: Blackwell.

Lenin, V. I. (1971) *Imperialism: The Highest Stage of Capitalism*, Peking: Foreign Languages Publishing House.

Levy, F. and Murnane, R. J. (1992) 'U.S. earnings levels and earnings inequality: a review of trends and proposed explanations', *Journal of Economic Literature* 43: 184–206.

Leyshon, A. (1992) 'The transformation of the regulatory order: regulating the global economy and environment', *Geoforum* 23: 249–67.

—— (1994) 'Under pressure: finance, geo-economic competition and the rise and fall of Japan's post-war growth economy', in S. Corbridge, R. Martin and N. Thrift (eds) *Money, Power and Space*, Oxford: Blackwell.

Leyshon, A. and Thrift, N. J. (1992) 'Liberalisation and consolidation: the single European market and the remaking of European financial capital', *Environment and Planning A* 24: 49–81.

Limes (1994) 'Le mie frontiere: colloquio di Rolf Gauffin con Vladimir Zirinovski', *Limes: Rivista Italiana di Geopolitica*, 2 (gennaio–marzo): 25–32.

Lipietz, A. (1987) *Mirages and Miracles: The Crises of Global Fordism*, London: Verso.

Lipsey, R. and Kravis, I. (1987) 'The competitiveness and comparative advantage of U.S. multinationals, 1957–84', *Banca Nazionale del Lavoro Quarterly Review* 161: 147–65.

Lipson, C. (1982) 'The transformation of trade', *International Organization* 36: 417–55.

—— (1989) 'International debt and national security: comparing Victorian Britain and postwar America', in B. Eichengreen and P. Lindert (eds) *The International Debt Crisis in Historical Perspective*, Cambridge MA: MIT Press.

—— (1991) 'Why are some international agreements informal?', *International Organization* 45: 495–538.

Lissakers, K. (1991) *Banks, Borrowers and the Establishment*, New York: Basic Books.

Little, I. (1982) *Economic Development*, New York: Basic Books.

Lloyd, P. J. (1992) 'Regionalisation and world trade', *OECD Economic Studies* 18: 7–43.

Lorenz, D. (1991) 'Regionalisation versus regionalism – problems of change in the world economy', *Intereconomics* January/February: 3–16.

Lowi, M. (1993) 'Bridging the divide: transboundary resource disputes and the case of West Bank water', *International Security* 18: 113–38.

Lucas, R. (1978) 'Unemployment policy', *American Economic Review* 68: 353–67.

Lugard, F. D. (1926) *The Dual Mandate in Tropical Africa*, Edinburgh: Oliver and Boyd.

Luke, T. W. (1991) 'The discipline of security studies and the codes of containment: learning from Kuwait', *Alternatives* 16: 315–44.

—— (1993) 'Discourses of disintegration, texts of transformation: re-reading realism in the new world order', *Alternatives* 18: 229–58.

Luttwak, E. (1982) *The Grand Strategy of the Soviet Union*, New York: St. Martin's Press.

—— (1990) 'From geopolitics to geo-economics', *The National Interest* 20: 17–24.

McAdams, A. J. (1990) 'Towards a new Germany? Problems of unification', *Government and Opposition* 25: 304–16.

MacEwan, A. (1986) 'Latin America: why not default?' *Monthly Review* 38: 1–13.

MacFarquhar, R. (1992) 'Deng's last campaign', *New York Review of Books* 39, 17 December: 22–8.

McHale, J. (1969) *The Future of the Future*, New York: George Braziller.

MacIntyre, A. (1981) *After Virtue: A Study in Moral Theory*, Notre Dame IN: University of Notre Dame Press.

—— (1989) *Whose Justice? Which Rationality?*, Notre Dame IN: University of Notre Dame Press.

Mackinder, H. J. (1904) 'The geographical pivot of history', *Geographical Journal* 13: 421–37.

McLuhan, M. (1966) *Understanding Media*, New York: Basic Books.

Maier, C. S. (1978) 'The politics of productivity: foundations of American international economic policy after World War II', in P. J. Katzenstein (ed.) *Between Power and Plenty: Foreign Economic Policies of Advanced Industrial States*, Madison WI: University of Wisconsin Press.

Makin, J. (1984) *The Global Debt Crisis: America's Growing Involvement*, New York: Basic Books.

Malmberg, B. (1992) 'A note on idealist models in social science', *Geografiska Annaler* 74 B: 117–23.

Mandrou, R. (1978) *From Humanism to Science, 1480–1700*, London: Penguin.

Mann, M. (1984) 'The autonomous power of the state: its origins, mechanisms and results', *European Journal of Sociology* 25: 185–213.

—— (1986) *The Sources of Social Power Vol. I*, Cambridge: Cambridge University Press.

Marglin, S. and Schor, J. (eds) (1990) *The Golden Age of Capitalism: Reinterpreting the Post-War Experience*, Oxford: Clarendon Press.

Markusen, A. R., Noponen, H. and Driessen, K. (1991) 'International trade, productivity, and U.S. regional job growth: a shift-share interpretation', *International Regional Science Review* 14: 15–39.

Marsh, D. (1992) *The Bundesbank: The Bank that Rules Europe*, London: Heinemann.

Marshall, M. (1987) *Long Waves of Regional Development*, New York: St. Martin's Press.

Martin, L. G. (1989) 'The graying of Japan', *Population Bulletin* 44.

Massey, D. (1984) *Spatial Divisions of Labour: Social Structures and the Geography of Production*, London: Methuen.

Mastanduno, M. (1988) 'Trade as a strategic weapon: American and Alliance export control policy in the early postwar period', *International Organization* 42: 121–50.

—— (1991) 'Do relative gains matter? America's response to Japan's industrial policy', *International Security* 16: 73–113.

Mazzolani, L. S. (1970) *The Idea of the City in Roman Thought: From Walled City to Spiritual Commonwealth*, Bloomington IN: Indiana University Press.

Mearsheimer, J. J. (1990) 'Back to the future: instability in Europe after the Cold War', *International Security* 15: 5–56.

Meinecke, F. (1972) *Historism: The Rise of a New Historical Outlook*, London: Routledge.

Miller, D. (1989) 'Why markets?', in J. LeGrand and S. Estrin (eds) *Market Socialism*, Oxford: Clarendon Press.

Milliken, J. L. (1990) 'Sovereignty and subjectivity in the early nineteenth century', Unpublished paper, Department of Political Science, University of Minnesota, Minneapolis MN.

Milner, H. (1988) *Resisting Protectionism: Global Industries and the Politics of International Trade*, Princeton NJ: Princeton University Press.

—— (1991) 'The assumption of anarchy in international relations theory: a critique', *Review of International Studies* 17: 67–85.

Minogue, K. (1990) 'Equality: a response', in G. Hunt (ed.) *Philosophy and Politics*, Cambridge: Cambridge University Press.

Mjoset, L. (1990) 'The turn of two centuries: a comparison of British and U.S. hegemonies', in D. P. Rapkin (ed.) *World Leadership and Hegemony*, Boulder CO: Lynne Rienner.

Mlinar, Z. (1992) 'Individuation and globalization: the transformation of territorial social organization', in Z. Mlinar (ed.) *Globalization and Territorial Identities*, Aldershot: Avebury.

Modelski, G. (1987) *Long Cycles in World Politics*, Seattle WA: University of Washington Press.

Moggridge, D. (1992) *Maynard Keynes: An Economist's Biography*, London: Routledge.

Mommsen, W. (1987) 'Flight from reality: Hitler as party leader and Dictator in the Third Reich', *Syracuse Scholar* 8, 1: 51–9.

Moore, B. Jr (1967) *Social Origins of Dictatorship and Democracy*, Boston MA: Beacon Press.

Moran, T. H. (1990) 'The globalization of America's defence industries: managing the threat of foreign dependence', *International Security* 15: 57–99.

Morgenthau, H. J. (1948) *Politics among Nations*, New York: Knopf.

Morgenthau, H. J. (1989) 'The danger of thinking conventionally about nuclear weapons', in C. Schaerf, B. Reid and D. Carlton (eds) *New Technologies and the Arms Race*, New York: St. Martin's Press.

Mosley, P., Harrigan, J. and Toye, J. (1992) *Aid and Power Vol. 1*, London: Routledge.

Moss, M. (1987) 'Telecommunications, world cities and urban policy', *Urban Studies* 24: 534–46.

Mosse, G. L. (1980) *Masses and Man: Nationalist and Fascist Perceptions of Reality*, Detroit MI: Wayne State University Press.

Mumford, L. (1961) *The City in History: Its Origins, Its Transformations, and Its Prospects*, New York: Harcourt Brace.

Murphy, C. N. (1994) *International Organization and Industrial Change: Global Governance since 1850*, Cambridge: Polity Press.

Murray, R. (1971) 'The intensification of capital and the nation-state', *New Left Review* 67: 84–109.

Nau, H. R. (1990) *The Myth of America's Decline: Leading the World Economy into the 1990s*, New York: Oxford University Press.

Neff, S. C. (1990) *Friends but No Allies: Economic Liberalism and the Law of Nations*, New York: Columbia University Press.

Newman, K. (1993) *Declining Fortunes: The Withering of the American Dream*, New York: Basic Books.

Nijman, J. (1992) 'The limits of superpower: the United States and the Soviet Union since World War II', *Annals of the Association of American Geographers* 82: 681–95.

Nolan, P. (1988) *The Political Economy of Collective Farms*, Cambridge: Polity Press.

Norris, C. (1992) *Uncritical Theory: Postmodernism, Intellectuals, and the Gulf War*, Amherst MA: University of Massachusetts Press.

Nozick, R. (1974) *Anarchy, State, and Utopia*, Oxford: Blackwell.

Nye, J. (1988) 'Neorealism and neoliberalism', *World Politics* 40: 235–51.

—— (1990a) *Bound to Lead: The Changing Nature of American Power*, New York: Basic Books.

—— (1990b) 'The changing nature of world power', *Political Science Quarterly* 105: 177–92.

O'Brien, P. K. and Pigman, G. A. (1992) 'Free trade, British hegemony and the international economic order in the nineteenth century', *Review of International Studies* 18: 89–113.

O'Brien, R. (1992) *Global Financial Integration: The End of Geography*, London: Pinter.

O'Connor, J. (1981) 'The fiscal crisis of the state revisited', *Kapitalistate* 9: 41–61.

Offer, A. (1993) 'The British Empire, 1870–1914: a waste of money?', *Economic History Review* 46: 215–38.

O'Loughlin, J. (1986) 'Spatial models of international conflict: extending current theories of war behavior', *Annals of the Association of American Geographers* 76: 63–80.

—— (1989) 'World-power competition and local conflicts in the Third World', in R. J. Johnston and P. J. Taylor (eds) *A World in Crisis?: Geographical Perspectives*, Oxford: Blackwell.

—— (1993) 'Fact or fiction? The evidence for the thesis of U.S. relative decline, 1966–1991', in C. H. Williams (ed.) *The Political Geography of the New World Order*, London: Belhaven Press.

O'Loughlin, J. and Grant, R. (1990) 'The political geography of presidential speeches, 1946–1987', *Annals of the Association of American Geographers* 80: 504–30.

O'Loughlin, J. and Van der Wusten, H. (1990) 'Political geography of panregions', *Geographical Review* 80: 1–20.

—— (1993) 'Political geography of war and peace', in P. Taylor (ed.) *The Political Geography of the Twentieth Century*, London: Belhaven Press.

Onuf, N. G. (1989) *World of Our Making: Rules and Rule in Social Theory and International Relations*, Columbia SC: University of South Carolina Press.

O'Tuathail, G. (1992a) 'The Bush Administration and the "end" of the Cold War: a critical geopolitics of U.S. foreign policy in 1989', *Geoforum* 23: 437–52.

—— (1992b) 'Putting Mackinder in his place: material transformations and myth', *Political Geography* 11: 100–18.

—— (1992c) '"Pearl Harbor without bombs": a critical geopolitics of the US–Japan "FSX" debate', *Environment and Planning A* 24: 975–94.

—— (1993a) 'The effacement of place? US foreign policy and the spatiality of the Gulf crisis', *Antipode* 25: 4–31.

—— (1993b) 'Japan as threat: geo-economic discourses on the US–Japan relationship in US civil society, 1987–1991', in C. H. Williams (ed.) *The Political Geography of the New World Order*, London: Belhaven Press.

O'Tuathail, G. and Agnew, J. A. (1992) 'Geopolitics and discourse: practical geopolitical reasoning in American foreign policy', *Political Geography* 11: 190–204.

243

Overbeek, H. and van der Pijl, K. (1993) 'Restructuring capital and restructuring hegemony: neo-liberalism and the unmaking of the post-war order', in H. Overbeek (ed.) *Restructuring Hegemony in the Global Political Economy: The Rise of Transnational Neo-liberalism in the 1980s*, London: Routledge.

Ozawa, I. (1994) *Blueprint for a New Japan*, New York: Kadansha International.

Padoa-Schioppa, T. (1993) *Tripolarism: Regional and Global Economic Cooperation*, Washington D.C.: Group of Thirty.

Palen, R. (1992) 'The European miracle of capital accumulation', *Political Geography* 11: 401–6.

Palma, G. (1978) 'Dependency: a formal theory of development or a methodology for the analysis of concrete situations of underdevelopment?' *World Development* 6: 881–924.

Parboni, R. (1984) *The Dollar and its Rivals*, London: Verso.

—— (1988) 'U.S. economic strategies against Western Europe: from Nixon to Reagan', *Geoforum* 19: 45–54.

Parfit, D. (1984) *Reasons and Persons*, Oxford: Oxford University Press.

Paterson, W. E. and Southern, D. (1991) *Governing Germany*, London: Norton.

Peet, R. (1989) 'The destruction of regional cultures', in R. Johnston and P. Taylor (eds) *A World in Crisis? Geographical Perspectives*, Oxford: Blackwell.

Perez, C. (1983) 'Structural change and assimilation of new technologies in economic–social systems', *Futures* October: 357–75.

Perkins, D. (1992) 'China's economic boom and the integration of the economies of East Asia', Unpublished paper, Department of Economics, Harvard University.

Phillips, K. (1990) *The Politics of Rich and Poor: Wealth and the American Electorate in the Reagan Aftermath*, New York: Random House.

Picciotto, S. (1991) 'The internationalization of the state', *Review of Radical Political Economics* 22: 28–44.

Pick, D. (1989) *Faces of Degeneration: A European Disorder, c. 1848 – c. 1918*, Cambridge: Cambridge University Press.

—— (1993) *War Machine: The Rationalisation of Slaughter in the Modern Age*, New Haven CT: Yale University Press.

Pipes, D. (1984) *Survival is Not Enough*, New York: Simon and Schuster.

Pletsch, C. E. (1981) 'The Three Worlds, or the division of social scientific labor, circa 1950–1975', *Comparative Studies in Society and History* 23: 565–90.

Pogge, T.W. (1993) 'Cosmopolitanism and sovereignty', *Ethics* 103: 48–75.

Polanyi, K. (1944) *The Great Transformation*, Boston MA: Beacon Press.

—— (1957) 'The place of economies in societies', in K. Polanyi (ed.) *Trade and Markets in the Early Empires*, Glencoe IL: Free Press.

Porter, M. (1990) *The Competitive Advantage of Nations*, London: Macmillan.

Poulantzas, N. (1980) *State, Power, Socialism*, London: Verso.

Pratt, J. W. (1935) 'The ideology of American expansion', in A. Craven (ed.) *Essays in Memory of William E. Dodd*, Chicago: University of Chicago Press.

Pratt, M. L. (1992) *Imperial Eyes: Travel Writing and Transculturation*, London: Routledge.

Ra'anan, U. (1991) 'Nation and state: Order out of chaos', in U. Ra'anan (ed.) *State and Nation in Multi-Ethnic Societies: The Breakup of Multinational States*, Manchester: Manchester University Press.

Randall, S. J. (ed.) (1992) *North America without Borders: Integrating Canada, the United States, and Mexico*, Calgary: University of Calgary Press.

Rao, V. (1993) 'Dowry "inflation" in India', *Population Studies* 47: 283–93.

Rasler, K. A. (1990) 'Spending, deficits, and welfare trade-offs: cause or effect of leadership decline?', in D. Rapkin (ed.) *World Leadership and Hegemony*, Boulder CO: Lynne Rienner.

Rasler, K. and Thompson, W. R. (1988) 'Defense burdens, capital formation, and economic growth', *Journal of Conflict Resolution* 32: 61–86.

Rawls, J. (1971) *A Theory of Justice*, Cambridge MA: Harvard University Press.

Reich, R. (1991a) *The Work of Nations: Preparing Ourselves for 21st Century Capitalism*, New York: Knopf.

—— (1991b) 'Who do we think they are?', *The American Prospect* 4: 49–53.

Riddell, J. (1987) *Foreign Aid Reconsidered*, London: James Curry.

Roberts, S. M. (1994) 'Fictitious capital, fictitious spaces? The geography of offshore financial flows', in S. Corbridge, R. Martin and N. Thrift (eds) *Money, Power and Space*, Oxford: Blackwell.

Roemer, J. (1988) *Free to Lose*, Cambridge MA: Harvard University Press.

Rokkan, S. and Urwin, D. (1983) *Economy, Territory, Identity: Politics of West European Peripheries*, London: Sage.

Rose, R. (1989) 'How exceptional is the American political economy?', *Political Science Quarterly* 104: 91–115.

Rosecrance, R. (1986) *The Rise of the Trading State: Commerce and Conquest in the Modern World*, New York: Basic Books.

Rosecrance, R. and Taw, J. (1990) 'Japan and the theory of leadership', *World Politics* 42: 184–209.

Rosenau, J. N. (1990) *Turbulence in World Politics: A Theory of Change and Continuity*, Princeton NJ: Princeton University Press.

Rosenberg, E. S. (1982) *Spreading the American Dream: American Economic and Cultural Expansion, 1890–1945*, New York: Hill and Wang.

Rosow, S. J. (1990) 'The forms of internationalization: representation of western culture on a global scale', *Alternatives* 15: 287–301.

Rostow, W. (1960) *The Stages of Economic Growth: A Non-Communist Manifesto*, London: Cambridge University Press.

Rotman, B. (1989) *Signifying Nothing: The Semiotics of Zero*, Basingstoke: Macmillan.

Routledge, P. (1993) *Terrains of Resistance: Nonviolent Social Movements and the Contestation of Place in India*, Westport CT: Praeger.

Rowen, H. S. and Wolf, C. (eds) (1990) *The Impoverished Superpower: Perestroika and the Soviet Military Burden*, San Francisco CA: Institute for Contemporary Studies.

Ruggie, J. G. (1983) 'International regimes, transactions and change: embedded liberalism in the postwar economic order', in S. D. Krasner (ed.), *International Regimes*, Ithaca NY: Cornell University Press.

—— (1993) 'Territoriality and beyond: problematizing modernity in international relations', *International Organization* 47: 139–74.

Rupert, M. E. (1990) 'Producing hegemony: State/society relations and the politics of productivity in the United States', *International Studies Quarterly* 34: 427–56.

Rupert, M. E. and Rapkin, D. P. (1985) 'The erosion of US leadership capabilities', in P. M. Johnson and W. R. Thompson (eds) *Rhythms in Politics and Economics*, New York: Praeger.

Rushdie, S. (1990) *Imaginary Homelands*, London: Granta.

Ryan, M. T. (1981) 'Assimilating New Worlds in the sixteenth and seventeenth centuries', *Comparative Studies in Society and History* 23: 519–38.

Sack, R. D. (1986) *Human Territoriality: Its Theory and History*, Cambridge: Cambridge University Press.

Sagan, C. and Turco, R. (1990) *Nuclear Winter and the End of the Arms Race*, London: Century.

Sahlins, P. (1989) *Boundaries: The Making of France and Spain in the Pyrenees*, Berkeley CA: University of California Press.

Said, E. W. (1978) *Orientalism*, New York: Vintage.

REFERENCES

—— (1979) 'Zionism from the standpoint of its victims', *Social Text* 1: 7–58.

—— (1988) 'Michel Foucault: 1924–1984', in J. Arac (ed.) *After Foucault*, New Brunswick NJ: Rutgers University Press.

—— (1993) *Culture and Imperialism*, New York: Knopf.

Samuels, R. J. (1989) 'Consuming for production: Japanese national security, nuclear fuel procurement, and the domestic economy', *International Organization* 43: 625–72.

Sanders, J. W. (1983) *Peddlers of Crisis: The Committee on the Present Danger*, Boston MA: South End Press.

Sandholtz, W. (1993) 'Choosing union: monetary politics and Maastricht', *International Organization* 47: 1–39.

Santos, M. (1979) *The Shared Space*, London: Methuen.

Schell, J. (1989) *Observing the Nixon Years*, New York: Random House.

Scheper-Hughes, N. (1992) *Death Without Weeping: The Violence of Everyday Life in Brazil*, Berkeley: University of California Press.

Schmiegelow, H. and Schmiegelow, M. (1989) *Strategic Pragmatism: Japanese Lessons in the Use of Economic Theory*, New York: Praeger.

—— (1990) 'How Japan affects the international system', *International Organization* 44: 553–88.

Schmitter, P. (1974) 'Still the century of corporatism?', *Review of Politics* 36: 85–131.

Schwoebel, R. (1967) *The Shadow of the Crescent: The Renaissance Image of the Turk (1453–1517)*, Nieuwkoop, The Netherlands: De Graaf.

Sen, A. (1978) 'Rational fools: a critique of the behavioral foundations of economic theory', in H. Harris (ed.) *Scientific Models and Men*, London: Oxford University Press.

—— (1989) 'Food and freedom', *World Development* 17: 769–81.

—— (1993) 'The economics of life and death', *Scientific American* May: 40–7.

Sender, J. and Smith, S. (1985) 'What's right with the Berg Report and what's left of its critic?', *Capital and Class* 24: 125–46.

Sheahan, J. (1980) 'Market-oriented economic policies and political repression in Latin America', *Economic Development and Cultural Change* 28: 267–91.

Shimko, K. (1992) 'Realism, neorealism and American liberalism', *Review of Politics* 54: 281–302.

Singh, I. (1990) *The Great Ascent: The Rural Poor in South Asia*, Washington D.C.: World Bank.

Slater, J. (1987) 'Dominos in Central America: Will they fall? Does it matter?', *International Security* 12: 105–34.

Smith, A. (1993) *Russia and the World Economy: Problems of Integration*, London: Routledge.

Smith, A. D. (1979) *Nationalism in the Twentieth Century*, Oxford: Martin Robertson.

—— (1991) *National Identity*, London: Penguin.

Smith, D. L. and Wanke, J. (1993) 'Completing the Single European Market: an analysis of the impact on the member states', *Journal of Politics* 37: 529–54.

Smith, W. D. (1980) 'Friedrich Ratzel and the origins of Lebensraum', *German Studies Review* 3: 51–68.

—— (1986) *The Ideological Origins of Nazi Imperialism*, New York: Oxford University Press.

Snidal, D. (1985) 'The limits of hegemonic stability theory', *International Organization* 39: 579–614.

Sommers, A. T. (1975) 'Social goals and economic growth: the policy problem in capital formation', *Conference Board Record* 12: 579–614.

Springborg, P. (1992) *Western Republicanism and the Oriental Prince*, Cambridge: Polity Press.

Spurr, D. (1993) *The Rhetoric of Empire: Colonial Discourse in Journalism, Travel Writing, and Imperial Administration*, Durham NC: Duke University Press.

Spykman, N. (1942) *America's Strategy in World Politics*, New York: Harcourt Brace.

Steele, J. (1994) 'Bear with a sore head', *The Guardian* 27 January: 2–3.

Stein, A. (1984) 'The hegemon's dilemma: Great Britain, the United States and the international economic order', *International Organization* 38: 355–86.

—— (1991) *Why Nations Cooperate: Circumstance and Choice in International Relations*, Ithaca NY: Cornell University Press.

Stockman, D. (1986) *The Triumph of Politics: The Inside Story of the Reagan Revolution*, New York: Harper and Row.

Stokes, E. (1959) *The English Utilitarians and India*, London: Oxford University Press.

Stopford, J. M. and Strange, S. (1991) *Rival States, Rival Firms: Competition for World Market Shares*, Cambridge: Cambridge University Press.

Strange, S. (1986) *Casino Capitalism*, Oxford: Blackwell.

—— (1987) 'The persistent myth of lost hegemony', *International Organization* 41: 551–74.

—— (1990) 'Finance, information and power', *Review of International Studies* 16: 259–74.

Strassoldo, R. (1992) 'Globalism and localism: theoretical reflections and some evidence', in Z. Mlinar (ed.) *Globalization and Territorial Identities*, Aldershot: Avebury.

Straszheim, D. H. (1991) 'Statement by Donald H. Straszheim, Chief Economist, Merrill Lynch & Co.', Joint Economic Committee, US Congress, Hearings on Midyear Economic Outlook, 26 July.

Taylor, P. J. (1993a) 'Geopolitical world orders', in P. J. Taylor (ed.) *The Political Geography of the Twentieth Century*, London: Belhaven Press.

—— (1993b) *Political Geography: World-Economy, Nation-State and Locality*, 3rd edn, London: Longman.

Thatcher, M. (1993) *The Downing Street Years*, London: Harper-Collins.

Thompson, J. A. (1992) 'The exaggeration of American vulnerability: the anatomy of a tradition', *Diplomatic History* 16: 23–43.

Thompson, W. R. (1990) 'Long waves, technological innovation, and relative decline', *International Organization* 44: 201–33.

Thomsen, S. and Woolcock, S. (1993) *Direct Investment and European Integration*, London: Pinter.

Thomson, J. E. (1994) *Mercenaries, Pirates, and Sovereigns: State-Building and Extraterritorial Violence in Early Modern Europe*, Princeton NJ: Princeton University Press.

Thrift, N. and Leyshon, A. (1988) '"The gambling propensity": banks, developing country debt exposure and the new international financial sysytem', *Geoforum* 19: 55–69.

Thrift, N. and Taylor, M. (1989) 'Battleships and cruisers', in D. Gregory and R. Walford (eds) *Horizons in Geography*, Basingstoke: Macmillan.

Thurow, L. (1985) *The Zero-Sum Solution*, New York: Simon and Schuster.

—— (1992) *Head to Head: The Coming Economic Battle Among Japan, Europe, and America*, New York: Warner.

Tillyard, E. M. W. (1943) *The Elizabethan World Picture*, London: Chatto and Windus.

Tomlinson, B. R. (1993) *The Economy of Modern India, 1860–1970*, Cambridge: Cambridge University Press.

247

Toulmin, S. (1990) *Cosmopolis: The Hidden Agenda of Modernity*, Chicago IL: University of Chicago Press.

Toye, J. (1993) *Dilemmas of Development: Reflections on the Counter-Revolution in Development Theory and Policy*, 2nd edn, Oxford: Blackwell.

Treverton, G. F. (1991/92) 'The new Europe', *Foreign Affairs* 71: 94–112.

Triffin, R. (1960) *Gold and the Dollar Crisis*, New Haven: Yale University Press.

Trubovitz, P. and Roberts, B. E. (1991) 'Bearing the burden: distributive politics and national security', Working Paper, Economics and National Security Program, Center for International Affairs, Harvard University.

Tylecote, A. (1992) *The Long Wave in the World Economy: The Current Crisis in Historical Perspective*, London: Routledge.

Tyson, L. and Zysman, J. (1989) *Politics and Productivity: The Real Story of Why Japan Works*, Cambridge MA: Ballinger.

U.S. Army (1944) *Geographical Foundations of National Power*, Washington DC: Army Service Forces, Manual M 103–1 – M 103–2.

van Creveld, M. (1991) *On Future War*, Oxford: Blackwell.

van der Pilj, K. (1984) *The Making of an Atlantic Ruling Class*, London: Verso.

van Wolferen, K. (1989) 'Japan: different, unprecedented and dangerous', *International Herald Tribune* 7 November: 6.

—— (1990) *The Enigma of Japanese Power: People and Politics in a Stateless Nation*, New York: Vintage.

Vernon, R. (1983) *Two Hungry Giants: The United States and Japan in the Quest for Oil and Ores*, Cambridge MA: Harvard University Press.

Virilio, P. (1986) *Speed and Politics*, New York: Semiotext(e).

—— (1989) *War and Cinema: The Logistics of Perception*, London: Verso.

Viroli, M. (1992) *From Politics to Reason of State: The Acquisition and Transformation of the Language of Politics, 1250–1600*, Cambridge: Cambridge University Press.

Volcker, P. and Gyohten, T. (1992) *The World's Money and the Threat to American Leadership*, New York: Times Books.

Wachtel, H. M. (1986) *The Money Mandarins: The Making of a New Supranational Economic Order*, New York: Pantheon.

Wade, R. (1990) *Governing the Market: Economic Theory and the Role of Government in East Asian Development*, Princeton NJ: Princeton University Press.

Wagner, R. H. (1993) 'What was bipolarity?', *International Organization* 47: 77–106.

Walker, R. B. J. (1990) 'Security, sovereignty, and the challenge of world politics', *Alternatives* 15: 3–27.

—— (1993) *Inside/Outside: International Relations as Political Theory*, Cambridge: Cambridge University Press.

Wallerstein, I. (1984) *The Politics of the World-Economy*, Cambridge: Cambridge University Press.

—— (1992) *Geopolitics and Geoculture*, Cambridge: Cambridge University Press.

Waltz, K. E. (1959) *Man, the State and War*, New York: Columbia University Press.

—— (1979) *Theory of International Politics*, New York: Random House.

—— (1990) 'Nuclear myths and political realities', *American Political Science Review* 84: 727–39.

Ward, M.D. (ed.) (1990) *The New Geopolitics*, London: Gordon and Breach.

Warf, B. (1989) 'Telecommunications and the globalization of financial services', *Professional Geographer* 41: 257.

Watts, M. (1989) 'The agrarian question in Africa: debating the crisis', *Progress in Human Geography* 13: 1–41.

—— (1991) 'Visions of excess: African development in an age of market idolatry', *Transition* 51: 124–41.

—— (1994) 'The devil's excrement: oil money and the spectacle of black gold', in S. Corbridge, R. Martin and N. Thrift (eds) *Money, Power and Space*, Oxford: Blackwell.

Webb, M. C. (1991) 'International economic structures, government interests, and international coordination of macroeconomic adjustment policies', *International Organization* 45: 309–42.

Weber, C. (1992) 'Reconsidering statehood: examining the sovereignty/intervention boundary', *Review of International Studies* 18: 199–216.

Weinstein, M. E. (1971) *Japan's Postwar Defense Policy, 1947–1968*, New York: Columbia University Press.

Welfens, P. J. J. (ed.) (1992) *Economic Aspects of German Unification: National and International Perspectives*, Berlin: Springer Verlag.

Wendt, A. (1992) 'Anarchy is what states make of it: the social construction of power politics', *International Organization* 46: 391–425.

Wicksteed, P. H. (1912) *The Common Sense of Political Economy Vol. II*, London: Routledge and Kegan Paul.

Williams, B. (1993) 'Freer than others: a review essay', *London Review of Books* 18 (November): 7–8.

Wills, G. (1992) *Lincoln at Gettysburg: The Words that Remade America*, New York: Simon and Schuster.

Wolf, E. R. (1982) *Europe and the People without History*, Berkeley CA: University of California Press.

Wolfe, A. (1981) *America's Impasse*, Boston MA: South End Press.

Wolin, R. (1993) *The Heidegger Contoversy: A Critical Reader*, Cambridge MA: MIT Press.

Wolin, S. S. (1960) *Politics and Vision: Continuity and Innovation in Western Political Thought*, Boston MA: Little, Brown.

—— (1989) *The Presence of the Past: Essays on the State and the Constitution*, Baltimore MA: The Johns Hopkins University Press.

Wood, R. (1986) *From Marshall Plan to Debt Crisis: Foreign Aid and Development Choices in the World Economy*, Berkeley CA: University of California Press.

Woodward, D. (1992) *Debt, Adjustment and Poverty in Developing Countries*, 2 volumes, London: Pinter.

World Bank (1978/81/85/87/92/93) *World Development Reports, 1978, 1981, 1985, 1987, 1992, 1993* New York: Oxford University Press.

—— (1981) *Accelerated Development in Sub-Saharan Africa*, Washington DC: World Bank.

—— (1989) *India: An Industrializing Country in Transition*, Washington DC: World Bank.

Wriston, W. (1986) *Risk and Other Four Letter Words*, New York: Harper and Row.

Yahil, L. (1990) *The Holocaust: The Fate of European Jewry*, New York: Oxford University Press.

Yates, P. L. (1959) *Forty Years of Foreign Trade*, London: Allen and Unwin.

Zacher, M. (1992) 'The decaying pillars of the Westphalian temple: implications for international order and governance', in J. N. Rosenau and E.-N. Czempiel (eds) *Governance without Government: Order and Change in World Politics*, Cambridge: Cambridge University Press.

Zerubavel, E. (1992) *Terra Cognita: The Mental Discovery of America*, New Brunswick NJ: Rutgers University Press.

INDEX